T. E. Thorpe

Qualitative Chemical Analysis and Laboratory Practice

T. E. Thorpe

Qualitative Chemical Analysis and Laboratory Practice

ISBN/EAN: 9783337139957

Hergestellt in Europa, USA, Kanada, Australien, Japan

Cover: Foto ©berggeist007 / pixelio.de

Weitere Bücher finden Sie auf **www.hansebooks.com**

TEXT-BOOKS OF SCIENCE

ADAPTED FOR THE USE OF

ARTISANS AND STUDENTS IN PUBLIC AND OTHER SCHOOLS.

QUALITATIVE CHEMICAL ANALYSIS

AND LABORATORY PRACTICE.

LONDON: PRINTED BY
SPOTTISWOODE AND CO., NEW-STREET SQUARE
AND PARLIAMENT STREET

Ba

QUALITATIVE

CHEMICAL ANALYSIS

AND

LABORATORY PRACTICE.

BY

T. E. THORPE, Ph.D. F.R.S.E.

PROFESSOR OF CHEMISTRY

ANDERSONIAN INSTITUTION, GLASGOW:

AND

M. M. PATTISON MUIR, F.R.S.E.

LONDON:

LONGMANS, GREEN, AND CO.

1874.

PREFACE.

THIS BOOK is divided into two distinct parts. In PART I. the student is instructed to perform a series of experiments, in order to familiarise himself with the leading properties of the chief non-metallic elements, and of the principal substances which they form by their mutual union. This portion of the book is, of course, supplementary to the work of the lecture-room, and should be studied in connexion with a manual in which the origin, properties, and relations of the various bodies are fully described.

The experiments are generally of a very simple nature, and are strictly illustrative of the chemical and physical properties of the substances to which they refer. To each lesson is appended a short statement or summary of the facts which that lesson is intended to convey. The object of these *résumés* is to afford the student precise ideas of the nature and extent of the information which he has gleaned from the experiments he has performed. The advantages of this preliminary course are manifest. Not only does the student become practically familiar with the properties of a large number of chemical agents, but he acquires opportunities for the exercise of manipulative skill and dexterity in the construction and arrangement of appa-

ratus, which the few and simple operations of ordinary
qualitative analysis are not so well fitted to give. In
arranging this course of laboratory practice we have sought
to cover the Scheme of Practical Instruction sketched
out in the Syllabus of the Science and Art Department.

PART II. treats of Qualitative Analysis ; it is divided
into five sections.

In the first section the general preliminary operations of
testing are described, such as the employment of flame
reactions, the use of the Bunsen flame, the spectroscope,
and so forth.

The second section treats of systematic qualitative
testing. The wet and dry reactions of each of the more
commonly occurring bases and acids, inorganic and organic,
are first described, after which a synopsis of analytical
methods is given.

The third section gives the special tests for the rarer
elements, and shows where they may be expected, and how
they may be separated, in the ordinary course of analysis.

The fourth and fifth sections have been added mainly
for the use of medical students.

We have endeavoured to make this book as practical as
possible. The methods of analysis are, of course, mainly
founded on established and reliable processes. Still the
book will be found to contain a number of novelties both in

the way of shortening the course of systematic testing and in the recognition of bodies by means of special tests. We have hesitated, however, to adopt new methods, unless experience has shown them to be preferable to the older ones. All the experiments described in the First Part have been carefully tried ; and with the exception of certain of the separations of the rarer metals and organic poisons, which, however, will be recognised as well-established processes, all the operations described in the Second Part have been repeatedly tested by ourselves and by students in this Institution.

Our thanks are due to Mr. DUGALD CLERK and to MR. STEPHEN MILLER for the attention they have bestowed on the woodcuts.

CONTENTS.

—◦—

PART I.

PREPARATION AND PROPERTIES OF GASES, LIQUIDS, AND SOLIDS.

LESSON I.

PAGE

Oxygen—Its Preparation—Bending Glass Tubes—Use of the Pneumatic Trough—Oxygen supports Combustion—Distinctive Character of Chemical Action—Meaning of the Term 'Chemical Test' 1

LESSON II.

Hydrogen—Its Preparation—Its Lightness—Pouring Hydrogen Upwards—Formation of Zinc Sulphate—Crystallisation of a Salt 9

LESSON III.

Combination of Oxygen with Hydrogen—Composition of Water—Combustion of Oxygen in Hydrogen—Combustion of Air in Coal Gas—Meaning of the term 'Combustion' . . . 12

LESSON IV.

Nitrogen – Its Preparation—It does not support Combustion—It is not itself combustible—It differs from Carbon Dioxide . . 18

LESSON V.

Nitric Acid—its Preparation—Distillation—Tests for Nitric Acid . 19

LESSON VI.

PAGE

Nitrogen Monoxide—Its Preparation—Meaning of the terms 'Neutralisation' and 'Salt'—Nitrogen Monoxide supports Combustion—Difference between its Power of supporting Combustion and that of Oxygen—Difference between Nitrogen Monoxide and a Mixture of Nitrogen and Oxygen 23

LESSON VII.

Nitrogen Dioxide—Its Preparation—Its Power of supporting Combustion—Difference between this and the Power of Nitrogen Monoxide—Action of Oxygen on Nitrogen Dioxide—Analysis of Air 27

LESSON VIII.

Ammonia—Its Preparation—Collection of Gases by Displacement —Combustion of Ammonia—Slow Combustion of Ammonia—Formation of Ammonium Chloride—Solubility of Ammonia in Water 30

LESSON IX.

Carbon Dioxide—Its Preparation—Its Density—Its Action on Lime Water and on Litmus Solution—It does not support Combustion—Decomposition of Carbon Dioxide—Its Solubility in Water—It is absorbed by Caustic Potash Solution . . . 36

LESSON X.

Carbon Monoxide—Its Preparation—Decomposition of Oxalic Acid by means of Sulphuric Acid—Separation of Carbon Monoxide and Dioxide—Combustion of Carbon Monoxide . . . 40

LESSON XI.

Chlorine—Its Preparation—Its Affinity for Hydrogen—Its Bleaching Action—Its Oxidising Action 43

LESSON XII.

PAGE

Hydrochloric Acid—Its Preparation—It does not support Combustion, nor is it Combustible—It is soluble in Water—Its Action on Litmus Solution—Synthesis of Hydrochloric Acid—Analysis of Hydrochloric Acid 47

LESSON XIII.

Bleaching Powder—Its Preparation—Theory of Bleaching—Hypochlorous Acid—Its Preparation—Potassium Chlorate—Its Preparation—Its Properties—Chlorine Tetroxide—Its Preparation 49

LESSON XIV.

Bromine—Its Preparation—Its Detection—Preparation of Potassium Bromide—Solubility of Bromine in Water—Combination of Bromine with Phosphorus 52

LESSON XV.

Iodine—Its Preparation—Sublimation—Tests for Iodine—Combination of Iodine with Phosphorus—With Sodium—Separation of Iodine from Bromine—Hydriodic Acid—Its Preparation—Its Properties—Difference between this Acid and Hydrochloric Acid 55

LESSON XVI.

Hydrofluoric Acid—Its Preparation—Etching Glass—Silicon Fluoride—Its Preparation—Its Properties—Hydrofluosilicic Acid—Its Preparation—Filtration—Use of the Wash Bottle . . 59

LESSON XVII.

Methane—Its Preparation—Its Combustion in Air—Products of its Combustion—Explosion of Mixture of Methane and Oxygen—Action of Chlorine upon Methane 65

Contents.

LESSON XVIII.

PAGE

Ethene—Its Preparation—Its Combustion in Air—Its complete Combustion with Oxygen—Its Action on Bromine and Chlorine 68

LESSON XIX.

Ethine—Its Preparation—Preparation of Cuprous Chloride— Detection of Ethine—Presence of Ethine in Coal Gas . . 71

LESSON XX.

Luminosity of Flame—Density of Products of Combustion—Zones of Flame—Extinction of Flame by Conduction of Heat by Metallic Surfaces 73

LESSON XXI.

Sulphur—Its Crystalline Forms—Test for Sulphur—Allotropy . 77

LESSON XXII.

Sulphur Dioxide—Its Preparation—Its Bleaching Action—Its Liquefaction—Its Oxidation—Its Action on Iodine—Sulphur Trioxide—Its Preparation—Its Properties 79

LESSON XXIII.

Sulphuric Acid—Its Properties—Its Action on Sugar . . . 85

LESSON XXIV.

Sulphuretted Hydrogen—Its Preparation—Its Action on Metallic Solutions 86

LESSON XXV.

Phosphoretted Hydrogen—Its Preparation—Its Inflammability . 88

PART II.

QUALITATIVE ANALYSIS.

SECTION I.

GENERAL PRELIMINARY OPERATIONS.

PAGE

Flame Reactions—Bunsen Lamp—Charcoal Splinter—Reduction and Oxidation of Substances—Films on Porcelain—Flame Colours—Coloured Glasses—Spectroscopic Analysis—Preliminary Dry Reactions 90

SECTION II.

SYSTEMATIC QUALITATIVE ANALYSIS.

Grouping of the Metallic Bases—Group Reagents—Special Tests for Members of Group I.—Separation of Group I. . . . 107
Special Tests for Members of Group II. and Separation of the Group 111
Special Tests for Members of Group III. and Separation of the Group 119
Special Tests for Members of Group IV. and Separation of the Group 122
Special Tests for Members of Group V. and Separation of the Group 126
Special Tests for Members of Group VI. and Separation of the Group 128
Illustration of Separation of all the Groups in the Analysis of Fahl Ore 130
Tests for Sulphuric Acid and Sulphates 134
,, ,, Hydrofluosilicic Acid and Silicofluorides . . 135
,, ,, Phosphoric Acid and Phosphates . . . 135
,, ,, Boric Acid and Borates 137
,, ,, Oxalic Acid and Oxalates 137
,, ,, Hydrofluoric Acid and Fluorides . . . 138
,, ,, Sulphurous Acid and Sulphites . . . 139
,, ,, Silicic Acid and Silicates 139
,, ,, Carbonic Acid and Carbonates . . . 141

PAGE

Tests for Iodic Acid and Iodates 141
,, ,, Thiosulphuric Acid and Thiosulphates 141
,, ,, Hydrochloric Acid and Chlorides 142
,, ,, Hydrobromic Acid and Bromides 142
,, ,, Hydriodic Acid and Iodides 143
,, ,, Chlorides, Iodides, and Bromides, when occurring to-
gether 145
,, ,, Hydrocyanic Acid and Cyanides 145
,, ,, Nitrous Acid and Nitrites 147
,, ,, Hypochlorous Acid and Hypochlorites 147
,, ,, Hydrosulphuric Acid and Sulphides 148
,, ,, Nitric Acid and Nitrates 149
,, ,, Nitrates and Chlorates, when occurring together . . 150
,, ,, Nitrogen in Organic Bodies 151
,, ,, Chloric Acid and Chlorates 151
,, ,, Perchloric Acid and Perchlorates 151
,, ,, Tartaric and Citric Acids 152
,, ,, Benzoic and Succinic Acids 153
,, ,, Acetic and Formic Acids 153

Synopsis of Analytical Methods :—

Table for Preliminary Dry Tests 155
,, ,, Flame Reactions by the Bunsen Flame . . . 156
,, ,, Solution of Substances 157
,, ,, Treatment of the Solution 158
,, ,, Separation of Group I. 159
,, ,, ,, ,, II. 160
,, ,, ,, ,, III. 161
,, ,, ,, ,, IV. 162
,, ,, ,, ,, V. 163
,, ,, ,, ,, VI. 164
,, ,, Preliminary Examination for Acids . . . 165
,, ,, Grouping of the Inorganic Acids . . . 166
,, ,, Separation of the Inorganic Acids . . . 167
,, ,, Examination of Insoluble Substances . . . 168
,, ,, Separation of Organic Acids 169

SECTION III.

DETECTION OF THE RARE ELEMENTS.

                                                         PAGE

Tests for Thallium 170

,, ,, Palladium, Rhodium, Osmium, Ruthenium, Platinum, and
        Iridium 171

,, ,, Molybdenum and Selenium 173

,, ,, Tellurium 173

,, ,, Zirconium, Cerium, Lanthanum, Didymium, and
        Titanium 174

,, ,, Uranium and Indium 176

,, ,, Vanadium, Lithium, Cæsium, and Rubidium . . . 177

Detection of the Rare Elements in the Systematic Course . . 178

SECTION IV.

DETECTION OF POISONS.

Detection of Phosphorus 182

,, ,, Arsenic by Dialysis 183

,, ,, ,, by Marsh's Process 187

,, ,, ,, by Fresenius and Von Babo's Process . . 192

,, ,, Mercury, Lead, and Copper 195

,, ,, Antimony and Zinc 195

,, ,, Hydrocyanic Acid 196

,, ,, Oxalic Acid 199

,, ,, Alkaloids by Method of Stas 200

,, ,, ,, ,, ,, Uslar and Erdmann . 202

Special Tests for Veratrine, Aconitine, Brucine, and Colchicine . 204

,, ,, ,, Morphine and Strychnine 206

,, ,, ,, Quinine 207

Detection of Opium 208

Identification of Blood Stains 209

SECTION V.

EXAMINATION OF URINE AND URINARY CALCULI.

Examination of Healthy Urine 210

,, ,, Abnormal Urine 210

,, ,, Urinary Calculi 214

APPENDICES 219

QUALITATIVE

CHEMICAL ANALYSIS

AND

LABORATORY PRACTICE.

———◆◇◆———

PART I.

PREPARATION AND PROPERTIES OF GASES, LIQUIDS, AND SOLIDS.

———

LESSON I.

PREPARATION AND PROPERTIES OF OXYGEN.

TAKE a few crystals of potassium chlorate, place them in a
clean dry test-tube, and heat them gently over a small Bunsen
flame (*see list of apparatus in Appendix*). The salt begins to
spirt, then fuses. Now light a small splint of wood, blow out
the flame so as to leave the wood just glowing at the point,
plunge this into the tube in which you are heating the potassium chlorate; but do not allow the wood to touch the fused
salt in the tube. Note what takes place: the splint, which
only glowed when introduced into the tube, now bursts into
bright flame, with a slight explosion. Withdraw the splint,
blow out the flame, introduce it again into the tube; the same
brilliant light will be noticed. What does this teach us?

B

That, by heating potassium chlorate, we have produced a
substance having the property of causing an almost expiring
flame to burn again more briskly than ever, which property,
as we can easily determine, the potassium chlorate does
not itself possess.   This substance is oxygen gas.   If
this oxygen has been produced from potassium chlorate,
then that salt must have undergone some change.   Let us
put this to the test.   Take a few crystals of the chlorate,
dissolve them in water in a test-tube (by shaking the tube,
the mouth being closed by the thumb), and to this solution
add a drop of a solution of silver nitrate.   Nothing seemingly takes place, the liquid remains clear.   Take now the
small quantity of potassium chlorate which you have heated
in the test-tube, dissolve this also in water, and add a drop
of silver nitrate solution ; at once a white solid forms in the
liquid, and this on shaking settles down to the bottom of the
tube, (such a solid substance, produced by the addition of
one solution to another, or, in some cases, by passing a gas
into a liquid, is called a *precipitate*).   These different actions
of silver nitrate show us that the potassium chlorate has been
changed by heat.   What we have after heating is potassium
chlorate *minus* oxygen, which, as we saw, goes off as a gas
($KClO_3 = KCl + O_3$).

In heating the potassium chlorate you will notice that it
spirts, and gives up moisture, which, condensing on the
colder part of the tube, is liable to flow back on to the hot
portion, and so to crack the glass.   Before preparing a quantity of oxygen it is therefore necessary to dry the salt.   To
do this, grind about 30 grams of potassium chlorate in a
mortar, place it in a porcelain dish which is supported on
a tripod over a very small flame, stirring it from time to time
with a glass rod until it is dry.   The temperature to which it
is necessary to heat potassium chlorate before it parts with
its oxygen is so high that the glass vessel containing the salt
is very apt to crack, or at least to soften ; but by mixing with
the chlorate a small quantity of an infusible substance, as

sand, oxygen will be given off at a comparatively low temperature. To facilitate the decomposition of the chlorate,
we generally use not sand, but manganese dioxide, which
acts more readily. Take about 5 grams ,of manganese
dioxide and set it drying, as you have done with the potassium chlorate.

Meanwhile get ready an apparatus in which to prepare
and collect the oxygen. Take a flask such as is described in
the list of apparatus, No. 1, clean and dry it; to do this,
pour a little water into the flask, add a few small pieces of
filtering paper, rinse well, shake out the paper, rinse again
with water, and let the flask drain; then heat it gently over a
flame, turning it round, and either suck or blow out the
heated air by means of a piece of glass tubing; select a
good cork a very little larger than the neck of the flask,
soften it (by wrapping it in a piece of paper, and rolling it
several times under the foot upon the floor), and bore a hole
in it about 6 mm. diameter, with a round file. (See Note 1.)
Choose now a piece of glass tubing about 85 cm. long and
very slightly wider than the hole in the cork. Light an
ordinary gas-burner, bring the tube at about 15 cm. from
one end into the flame, holding it parallel with the broadest
part (fig. 1), and keep turning it round; soon you will feel
the tube soften. Now
slowly bend it, as if you
were making a crook
like a shepherd's staff,
turning it in the flame
as much as possible, and
not hurrying the opera-

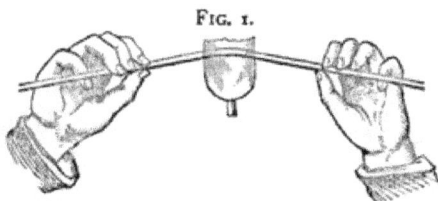

FIG. 1.

tion in the least. Some practice is required to make a neat
bend; but if the flame be good, if you turn the tube inces-
santly, and if you bend it *slowly*, you will soon be able to
conquer any little difficulty which at first may present itself.
When bent, the tube should have the shape shown in fig. 2,
No. 1. At about 15 cm. or so from the other end make

another bend in the opposite direction, but do not make this
bend so sharp as the first (fig. 2, No. 2). In making this
bend, hold the tube so that you may look along it, and then
you will be able to make both bends in one plane. Now,
by means of a three-cornered file, make a scratch on the
tube about 10 cm. from the end (fig. 2, No. 2, c), and, by a
sharp pull, break the tube at this point. Holding the tube
in both hands, bring one end of it into a Bunsen's flame, the
tube being inclined at an acute angle with the flame (see fig.
2, No. 1) ; keep turning it round until just the edge of the
tube begins to glow, then after a few moments (turning the
tube all the while) withdraw it, when you will find that the

FIG. 2.

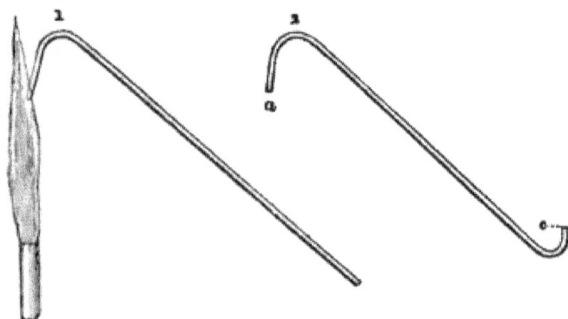

rough edge is now perfectly smooth ; do the same with the
other end of the tube, and allow it to cool. The end *a*
(see fig. 2) is now to be gently fitted into the cork. By this
time the potassium chlorate and manganese dioxide will be
dry; these are mixed together in a mortar, the mixture placed
upon a piece of paper, and poured, by means of this, into the
flask. Fit the cork with the bent tube into the flask and
support it by a clamp (see fig. 3).

To collect the gas, use the pneumatic trough and four
gas bottles. The trough is nearly filled with water, in
which a gas bottle is inverted so that the water may rush
into it, the air escaping as the water enters. Some little

management is required to fill the bottle completely, but
by raising the mouth—the bottle lying on its side beneath
the water—just above the surface, and then slowly depressing it, the last bubble of air may be forced out.
When this is done, place the bottle mouth downwards to
one side of the trough, while you fill the other bottles in
the same way.   Now place the flask on the retort-stand in
such a position that the lower end of the delivery tube
passes under the water just beneath the hole in the beehive

Fig. 3.

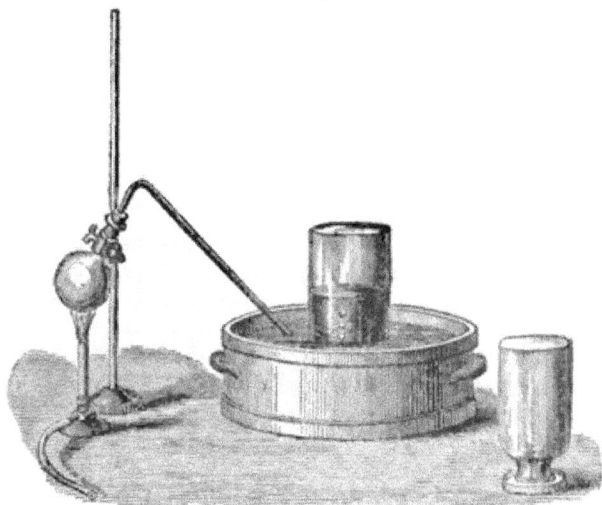

shelf of the trough.   Everything being now ready (see fig. 3),
gently heat the flask, using at first a small flame ; the air in
the flask will be first of all expanded and driven out, bubbling up through the water in the trough.   When the bubbles
begin to follow one another in rapid succession they may
be tested by bringing a glowing splint of wood over the
place where they stream up through the water.   If the splint
bursts into flame, then we may begin to collect the gas.   To
do this, move one of the bottles on to the beehive shelf, so
that its mouth is over the opening in this shelf ; the gas will

then rise into the bottle, displacing the water, which finds its way into the trough. When full, depress the mouth of the bottle somewhat further beneath the water, slip a small tray under it, then lift the bottle, standing mouth downwards, on the tray (which must contain a little water) out of the trough, and set it aside.

Fill the other three bottles in a similar manner. In heating the flask while preparing the gas, so regulate the heat as to avoid any sudden rush of gas ; and whenever a sufficient amount of oxygen has been collected, remove the lamp and lift the end of the delivery tube out of the water, otherwise, in cooling, the water will rush back into the flask and crack it. The mass remaining in the flask may, when cold, be easily washed out with water.

We will now proceed to examine the properties of the oxygen we have obtained. By the preliminary experiment with potassium chlorate heated in the test-tube we have learned that a burning body when plunged into oxygen burns with greatly increased activity. *Oxygen is evidently a supporter of combustion.*

Experiment I.—Take a small piece of phosphorus about tne size of a pea, dry it carefully between folds of filtering paper (see Note 2), and place it in the cup of a deflagrating spoon, set fire to the phosphorus by bringing it for a moment into a gas flame, and plunge it into a bottle of oxygen ; a brilliant white light, almost insupportably bright, is produced, together with dense white fumes. (If the gas bottle be made of thick glass, it is advisable to dilute the oxygen in it with about one-third of its volume of air before burning the phosphorus, as the great heat may crack the bottle. This may be done by depressing the bottle in the trough, allowing one-third of the oxygen to escape, and then lifting the bottle, mouth downwards, out of the trough, so that the water which flowed in when the oxygen escaped, may run out.) When the bottle is cool, remove the deflagrating spoon, put a little water into the bottle, and shake it up; the white fumes disappear, they are dissolved in the water. Taste a few drops of the solution thus formed; you find it exceedingly sour or *acid*. Pour into the bottle a few drops of blue litmus solution (reagents list, No. 27); it turns red, showing the presence of an acid. What, then, has this experiment taught us?

(1) That substances burn in oxygen with great energy.

(2) That a substance having totally different properties from either oxygen or phosphorus is produced when these bodies combine together. The colourless gas oxygen, and the yellow wax-like solid phosphorous form, when chemically combined, a light snow-like mass, which, when dissolved in water, gives rise to a solution possessing an intensely sour taste.

Experiment II.—Place a small piece of sulphur in the deflagrating spoon, set fire to it, and bring it into a bottle of *undiluted* oxygen; it burns with a pale lavender-blue flame, considerably more brilliant than when it burns in air. After the combustion is over, remove the spoon, and note the pungent odour of the substance formed, also the seeming absence of anything in the bottle; the product of combustion is an invisible gas. Add a little water, shake the bottle, and pour in a drop or two of blue litmus solution. By its action on the litmus you see that an acid is present.

Experiment III.—In the third bottle a small piece of charcoal is burned, in the spoon. Note the comparatively feeble light produced; the presence of the gaseous compound formed is shown (1) by its extinguishing a lighted taper when plunged into it; (2) by its action on lime-water, which, when added and the bottle shaken, becomes turbid, owing to the gas in the bottle, known as carbon dioxide,

CO₂, combining with the lime (lime-water being a solution of lime in water), and forming a substance, carbonate of lime, or calcium carbonate, insoluble in water, and therefore appearing as a solid, floating about in small particles in the clear liquid, and thus rendering it turbid.

You have thus detected the presence of an invisible gas, by presenting to it something with which it could combine to form a new substance, having such characteristic properties as to be easily recognisable; you have thus applied a *chemical test.*

You have learned from these experiments :—

(1) That oxygen supports combustion.

(2) That the production of substances differing in properties from the substances which combine to form them is a characteristic result of *chemical action.*

(3) The meaning of the term *chemical test.*

The fourth bottle of oxygen is set aside, with its mouth downwards, in one of the small trays · we shall have occasion to use it in Lesson IV.

Note 1.—In boring a hole in a cork, do not forcibly drive the end of the file into the cork, but gently turn the file round, at the same time pressing it cautiously into the cork, until the hole is made; take care that the hole be bored straight through, and not in a slanting direction. If you wish to enlarge the hole, you may easily do so by working the file round and round, filing away thin little bits of cork; but see that you do this equally on all sides, otherwise the round glass tube will not perfectly fit the hole in the cork. Cork-borers, which are sold in sets, are very useful, making a neater hole than a file.

Note 2.—In working with phosphorus, be careful to keep it under water. To prepare a piece for Experiment I., take a little piece out of the bottle by means of a knife; place it in a basin under water, and then cut off a piece of the required size; dry this by pressing it several times *quickly* between folds of blotting or filtering paper.

LESSON II.

PREPARATION AND PROPERTIES OF HYDROGEN.

TAKE one or two grams of zinc, put them into a test-tube, and add a few drops of dilute sulphuric acid ; you notice an effervescence, bubbles of gas make their way through the liquid ; bring a lighted taper into the test-tube : a slight explosion takes place, and you see a momentary flash, as if something in the tube had taken fire.

FIG. 4.

You have here another instance of chemical action : the zinc has so acted on the sulphuric acid as to decompose or break it up, hydrogen gas, one of the constituents of the acid, being evolved. You must now proceed to prepare and collect a quantity of hydrogen, making use of the reaction you have just learned for this purpose.

Take a flat-bottomed flask (similar to that represented in fig. 4), holding about 300 c.c.; select a cork, soften it, and bore in it two holes of such a size as to fit the bent tube you used in last lesson. In boring the holes, take care you do not make them too near the edges of the cork, nor too near one another, otherwise the cork will be sure to crack when you attempt to fit it into the mouth of the flask. Into one of these holes you fit the bent delivery tube, and into the other a straight glass tube, terminating at its upper end in an open mouth-shaped funnel ; this funnel-tube is to be pushed through the cork, so that when fitted into the flask it reaches very nearly to the bottom. Take now about 30 grams of granulated zinc (see reagents list), put it into the flask, inclining the flask to one side, and gently sliding the zinc down the neck, taking care that it does not fall heavily against the bottom, else it will most likely crack the flask; fit the cork into its place, and arrange the apparatus so that the delivery tube may reach under the shelf of the trough (fig. 4).

Pour into the flask through the funnel-tube sufficient water to cover the zinc to the depth of about 6 mm. or so ; then add, also through the funnel, a small quantity of diluted sulphuric acid (one part acid to four water, previously mixed and allowed to cool), and shake the flask gently ; in a minute or two you will notice an effervescence ; gas begins to bubble up through the water in the trough ; after a few moments add a little more acid, so that a pretty rapid stream of gas may be obtained. Do not, however, add much acid at a time, otherwise the action will become violent, the liquid will get very hot, and will froth up through the funnel-tube ; should there be indication of this, pour a little water into, and also on to the outside of the flask, in order to cool it. Take a test-tube, fill it with water in the trough, and bring it, mouth downwards, over the hole in the shelf whence the gas is issuing ; when it is full of gas, cover the mouth of it with the thumb (while still in the trough), lift it mouth downwards out of the water, remove the thumb, and bring a lighted taper to the mouth of the tube ; if an explosion ensues, the hydrogen is not pure, it is still mixed with air ; repeat this experiment after the expiration of a minute or so, and do not begin collecting the gas in bottles until a small quantity in a test-tube does not explode when brought near a light, but takes fire and burns quietly. When you have satisfied yourself that the hydrogen is unmixed with air, proceed to collect four bottles of it, exactly as described in the last lesson.

Be careful never to bring a light near a mixture of hydrogen and air in a glass vessel, as if these gases be present in any quantity, and the vessel is not very strong, a dangerous explosion may occur.

Let us now examine some of the properties of this gas hydrogen, and see what lessons it has to teach us.

Experiment I.—Take a bottle of hydrogen from the little tray, hold it mouth downwards, and apply a light to the mouth of the jar ; the gas takes fire, burning with a very pale non-luminous flame ; pass the taper further up into the jar, it is extinguished. Hydrogen therefore differs from oxygen in being itself combustible, but not (under ordinary circumstances) a supporter of combustion. You may naturally inquire, Why keep the bottle of hydrogen inverted? The next experiment will answer this question.

Experiment II.—Take the second bottle of hydrogen, lift it from the tray, keeping it inverted, in the right hand, while in the left you hold, also mouth downwards, the bottle in which you performed Experiment I., containing now only air ; by depressing the top of the bottle containing hydrogen, and holding the empty bottle as if you were pouring from the lower one into it (see fig. 5), you can in a very few moments decant all the hydrogen upwards into the higher bottle. Set the bottle from which you have poured the hydrogen, mouth downwards, in the tray ; light a taper and bring it to the mouth of the second bottle, still held inverted in the left hand : a slight explosion ensues, and you see the pale lambent flame of burning hydrogen. Apply

FIG. 5.

a light now to the mouth of the other bottle—the taper burns quietly, as if in air, showing you that all the hydrogen has been poured out of this bottle. You learn then that hydrogen is much lighter than air, quickly rising through it. Had you held the bottle in Experiment I. mouth upwards, the light hydrogen would have escaped before you could notice its characteristic flame.

Pour the liquid from the flask which you have used in these experiments into a small porcelain basin, evaporate it over a flame to about one-half or one-third its bulk, and set it aside ; when cool, you will find a mass of white crystals formed in the liquid ; this is zinc sulphate, the second product of the action of zinc on sulphuric acid, the first being the hydrogen you collected. The action is thus represented :

$$Zn + H_2SO_4 = ZnSO_4 + H_2.$$

You have learned from these experiments—

(1) That hydrogen is a combustible gas, burning with a feebly luminous flame.

(2) That hydrogen is an exceedingly light substance, so light that it can be poured *upwards*.

(3) That hydrogen and air form an explosive mixture.

(4) That the action of zinc on sulphuric acid is to produce an invisible gas, hydrogen, and a white crystalline solid, zinc sulphate.

LESSON III.

COMBINATION OF OXYGEN WITH HYDROGEN.—MEANING OF THE TERMS 'COMBUSTIBLE' AND 'SUPPORTER OF COMBUSTION.'

Experiment I.—Procure an ordinary soda-water bottle and a cork to fit it tightly ; fill it with water, pour out this water into a measuring glass—you thus find the capacity of the bottle ; divide this by three, and now pour into the bottle one-third of the water which you know is required to fill it. With a file make a mark on the outside of the bottle on a level with the surface of the water inside. You have thus divided the bottle into two parts : one equal to one-third, the other equal to two-thirds of the whole. Again fill the bottle with water at the trough, invert it, and let it stand on the beehive shelf, with its mouth under water. You now bring to the side of the trough the bottle of oxygen which you prepared but did not use in Lesson I., also a bottle of hydrogen from last lesson. Lift the bottle of oxygen, with the tray on which it stands, into the trough ; when the

FIG. 6.

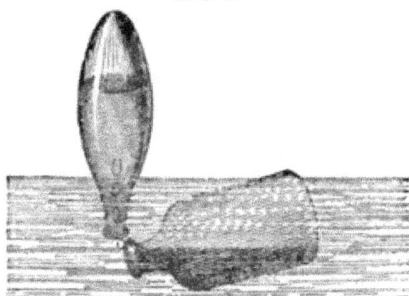

mouth of the bottle is beneath the surface of the water, with-draw the tray, gently depress the upper end of the bottle (held in the right hand), while with the other hand you hold the soda-water bottle so that the bubbles of oxygen as they escape may pass up through the water into it (see fig. 6). A little dexterity is necessary in thus decanting the gas up through the narrow neck of the soda-water bottle ; do not be too hasty, but allow the gas to pass up slowly and steadily. Fill the soda-water bottle with oxygen up to the mark you made with the file, that is, one-third ; then set it on the beehive shelf while you again bring the tray under the mouth of the oxygen bottle and lift it from the trough. Decant (as you have just done with oxygen) hydrogen sufficient to entirely fill the soda-water bottle, and set aside

any hydrogen which remains.   Quickly cork the soda-
water bottle, shake it once or twice briskly, withdraw the
cork, and apply a light to the mouth of the bottle—a sharp,
loud explosion instantly ensues.   With the oxygen and
hydrogen remaining, again fill the bottle, but this time vary
the proportion of the gases—put less oxygen and more
hydrogen, or *vice versa* ; you will find that the explosion is
less violent than in the first experiment.

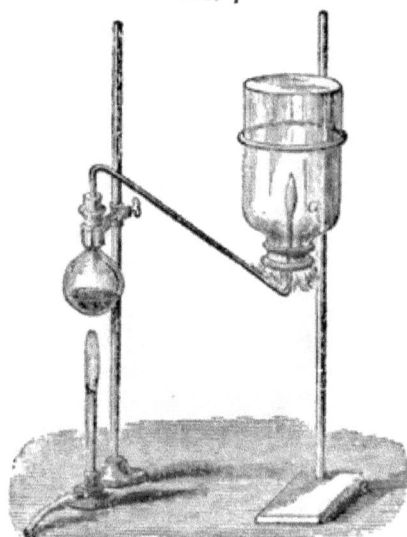

FIG. 7.

The application of the light has caused chemical combination to take place between the two gases, oxygen and
hydrogen ; this combination is of a very violent nature, as
testified by the loudness of the explosion.   The product of
the combination is water, which is composed cf oxygen and
hydrogen in the proportion of one volume of the former gas
to two volumes of the latter.   From the small volume of the
gases employed, the quantity of water produced is so minute
that it seemingly adds nothing to the drops already adhering
to the inside of the soda-water bottle.   If, however, we were
to fill a very large and perfectly dry bottle with the two gases
mixed in the above proportion and explode the mixture, we

should notice that a perceptible quantity of water formed as dew on the sides of the bottle.

With the fourth bottle of hydrogen prepared in last lesson you have another experiment to perform.

Experiment II.—Set up again the flask with bent delivery tube for the preparation of oxygen, putting into it a mixture of about 15 grams of potassium chlorate, and 3 grams of manganese dioxide, as directed in Lesson I. Soften the end, fig. 7, of the delivery tube in the flame, and keep turning it round until the edges run together, and so make the orifice much narrower than it originally was. Support the flask on a stand, so that the end of the delivery tube points straight upwards, and is about 20 cm. from the table. (No trough is needed in this experiment.)

Heat the flask gently, so as to produce a slow current of oxygen gas; from time to time bring a smouldering splint of wood near the end of the tube, and when the nearly expiring flame just bursts again into brilliancy, you may consider the stream of oxygen to be issuing with sufficient rapidity. Now lift the bottle of hydrogen, mouth downwards, from the tray; apply a light to it, and while the hydrogen is burning, place the mouth of the jar over the end of the tube from which oxygen is issuing. As you do this, you notice that at the point where the oxygen meets the burning hydrogen a long flame of ignited oxygen shoots up within the bottle (fig. 7).

If the oxygen is issuing in a rapid stream, there is danger of the flame produced by it being so large as to strike against the upper part of the bottle and crack it. You must, therefore, carefully regulate the supply of oxygen.

In this experiment you see that a gas which you had previously regarded as a supporter of combustion can itself be made to undergo combustion. As combustion is a *chemical combination* attended with evolution of light and heat, this action takes place where the two substances which combine meet each other; but the oxygen rushes into an atmosphere of hydrogen, surrounding it on all sides; the particles of oxygen,

as they are combining with those of hydrogen, are carried
quickly forwards, and thus it is that the oxygen appears itself
actually to burn within the hydrogen with the long narrow
flame which you noticed.

The relativity of the terms ' combustion ' and ' supporter
of combustion ' may be shown with other gases than pure
oxygen and hydrogen.   Thus air may be made to burn in
an atmosphere of coal gas.

For this experiment you require a flask with three necks.
To make this, take a round-bottomed flask, of about 300 c.c.
capacity (a Florence flask suits admirably), and at the points
a and *b* (fig. 8, No. 1) heat it in the blowpipe flame, at first
very gently, then gradually increasing the heat ; use a small

Fig. 8.

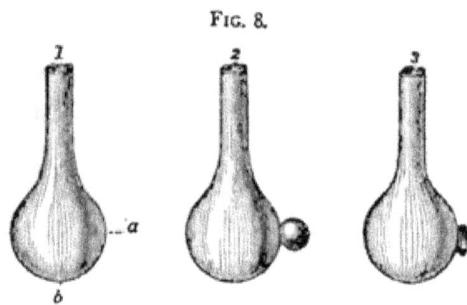

flame, and direct it so as to heat to redness a small circle of
glass at the point *a* ; when this is soft, suddenly blow in at
the mouth of the flask, the glass at *a* is blown out (see fig. 8, 2)
and breaks.   Now, still using the blowpipe flame, fuse the
rough edges of the opening thus made ; these melt and run
together, and you obtain a round hole with smooth edges
(fig. 8, 3) ; repeat this operation at the point *b*, a cork being
fitted into the hole *a* while you blow out the glass.

You have now a flask with three necks.

Choose a cork to fit, not very tightly, into the lower neck
of the flask, and through this cork pass a piece of glass tubing
about 10 cm. long, drawn to a tolerably fine point at its
upper end.

Fit a cork also into the side neck of the flask, with a glass tube passing through it, and arrange the two tubes so that the end of the one inside the flask may be just over that of the other (fig. 9). To the tube entering the flask at the side attach a caoutchouc tube coming from the gas supply, turn on the gas, and after the expiration of a minute or two light it as it issues from the upper neck of the flask; now cautiously draw out the lower cork with its tube, and when it is just out of the flask, apply a light to the neck; the gas will take fire there; as soon as it does so fit the cork into its place, and you will perceive a small pale flame playing round the point of the tube. The formation of this flame is accounted for by a similar reasoning to that used in explaining the burning of oxygen in hydrogen; in this case air takes the place of oxygen, and the atmosphere of hydrogen is replaced by one of coal-gas.

FIG. 9.

In this lesson you have learned :—

(1) That oxygen and hydrogen unite together violently when brought near a flame.

(2) That the union of two volumes of hydrogen with one volume of oxygen gives a louder explosion than any other proportionate mixture of these two gases.

(3) That water is composed of oxygen and hydrogen in the above proportion, viz., two volumes of hydrogen and one volume of oxygen.

(4) That the terms 'supporter of combustion,' and 'combustile,' are merely relative, and denote no absolute property of bodies.

LESSON IV.

PREPARATION AND PROPERTIES OF NITROGEN.

THE gas nitrogen exists already made in the air ; in order
to obtain it, you have only to take away another gas—oxygen
—with which it is mixed.

Fill the trough to the depth of about 3 cm. with water,
and place in it an ordinary white plate. Into this plate
pour a little water coloured blue by the addition of litmus
solution. (See fig. 10.)

Float a small porcelain dish, about 5 cm. diameter, on the
water :—the lid of a porcelain crucible, fastened to a piece

FIG. 10.

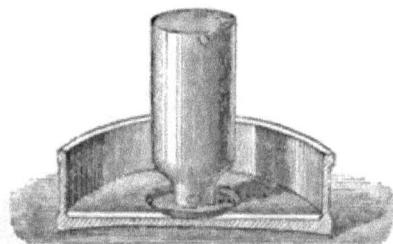

of wood of sufficient size to
float it, does very well. Cut
a little piece of phosphorus,
attending to the directions
given in Lesson I., dry it,
place it on the dish, and set
fire to it by touching it with
a hot wire. Have at hand
a large stoppered bottle, holding about 750 c.c. Take out
the stopper, and place the bottle cautiously over the dish
containing the burning phosphorus. You notice that as
the phosphorus burns, the water rises slightly inside the
bottle, showing that something is being taken away from
the enclosed air. That it is the oxygen which is thus taken
up by the phosphorus, is evident from the fact that the
blue colour of the water gradually changes to red. This,
you remember (Lesson I.), is owing to the formation of an
acid, the product of the union of phosphorus and oxygen.
When the combustion is over, and the bottle is cool, fill up
the trough with water (take care that the bottle does not

upset as you do this), and decant the air or gas into two gas bottles (see Lesson III.).

Experiment I.—Plunge a lighted taper into one of the bottles; the flame is at once extinguished, and the gas does not take fire. Nitrogen therefore is incombustible, and is a non-supporter of combustion.

Experiment II.—Into the other bottle pour a few cubic centimetres of clear lime-water, close the bottle with the palm of the hand, and briskly agitate it. The lime-water is not rendered turbid. This test serves to distinguish nitrogen from carbon dioxide, which, as you learned from Lesson I., extinguishes flame, but gives a precipitate with lime-water.

You learn then—

(1) That air contains oxygen and nitrogen.

(2) That by burning a combustible substance in a confined space of air, the oxygen is withdrawn (the combustible uniting with it), and nitrogen remains.

(3) That this nitrogen possesses no very distinctive qualities, being neither combustible, nor supporting combustion.

LESSON V.

PREPARATION AND PROPERTIES OF NITRIC ACID.

CHOOSE a stoppered retort, capable of holding about 250 c.c., having a tolerably long beak; clean and dry it. Weigh out 30 grams of potassium nitrate, place it on a small piece of paper turned up at the edges, so as to form a kind of little trough, and by means of this transfer it to the retort, taking care that none of it gets into the neck of the retort (see fig. 11). We act on this nitre with sulphuric acid; the equation expressing this reaction is—

C 2

$KNO_3 + H_2SO_4 = HNO_3 + KHSO_4$—that is to say,

$39 + 14 + 48 = 101$ parts by weight of potassium nitrate are decomposed by—

$2 + 32 + 64 = 98$ parts by weight of sulphuric acid; the product being—

$1 + 14 + 48 = 63$ parts by weight of nitric acid.

Calculate then the amount of strong sulphuric acid required to decompose the 30 grams of nitre, weigh out this amount in a small beaker, and pour it into the retort through a small funnel passing well into the body of the retort through the opening at the neck. Put in the stopper and support the retort, with wire gauze beneath it, on one ring of a retort-stand, bringing another ring over the neck. Clean and dry a small flask, the neck of which will permit the

FIG. 11.

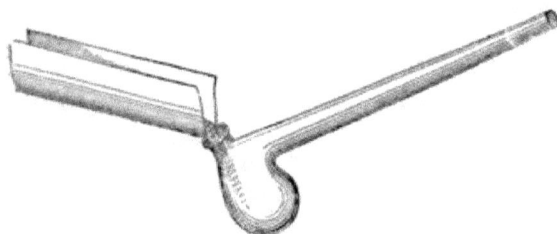

beak of the retort to be passed through it; support it on the beehive shelf, and fill the trough with water, so that the flask may be pretty well covered, but take care that no water gets into it.

The apparatus has now the appearance shown in Fig. 12. Gently heat the retort: an action evidently goes forward; after a little time brown fumes are evolved, and drops of liquid form on the sides of the neck, and trickle slowly down it into the small flask, or *receiver*, which being kept cool by the water surrounding it, condenses the nitric acid, which in the heated retort is in a gaseous state.

In the receiver a brownish-yellow heavy liquid, fuming

in the air, gradually accumulates ; this is strong nitric acid, which we will now examine.

When you have removed the light, pour out the semi-liquid contents of the retort, consisting chiefly of hydrogen potassium sulphate ($HKSO_4$), into a small basin. After cooling and solidifying, break up this salt into little pieces, and preserve it in a stoppered bottle.

You already know what is meant by a *chemical test*. In the laboratory we often have occasion to detect this substance nitric acid, and to do this we use certain chemical tests.

Experiment I.—Pour a little of the acid into a test-tube

FIG. 12.

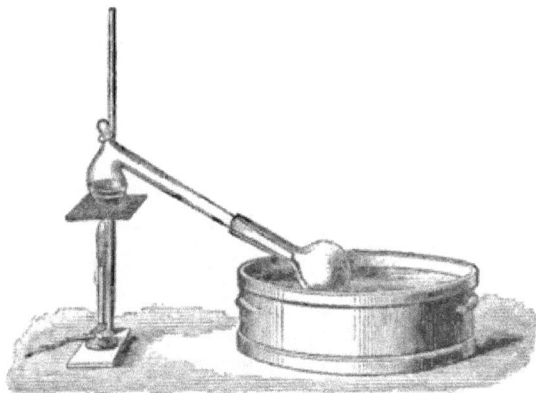

add a few drops of water, and then put into the tube a little slip of copper foil ; you notice that a violent action ensues, a dark reddish-brown gas is given off, and the liquid becomes more or less green in colour.

The production of these red fumes when a piece of copper foil is brought into contact with a liquid suspected to be nitric acid, is one proof that the liquid *is* nitric acid.

The red-coloured gas consists chiefly of nitrogen tetroxide, NO_2—

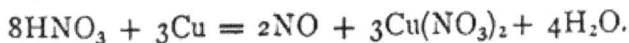

$$8HNO_3 + 3Cu = 2NO + 3Cu(NO_3)_2 + 4H_2O.$$

The NO thus formed, in contact with air takes up

another atom of oxygen, forming the ruddy-coloured gas
NO_2.   But we have by the production of these same ruddy
fumes, and the application to them of another chemical test,
a much more delicate means of detecting the presence of
small quantities of nitric acid.

Experiment II.—Take a few crystals of *ferrous sulphate*
and dissolve them in a little *cold* water, in a test-tube.   Into

FIG. 13.

the test-tube used in Experiment I. pour
a few drops of nitric acid, so as to cause
a fresh evolution of gas.   Now hold the
two tubes so that the ruddy fumes may
fall into that containing the ferrous sulphate solution (see fig. 13); the clear solution becomes
rapidly darkened in colour, until it is nearly black; this
darkening effect of the oxides of nitrogen, when brought
into contact with ferrous sulphate solution *in the cold*, is
the second and most delicate test for the presence of nitric
acid.   (Boil the dark brown solution in ferrous sulphate;
you see that the colour is almost entirely discharged.   You
now perceive why directions have been given to apply this
test in cold solutions.)

To carry out the test—

Experiment III.—Make a solution of nitre in a little
water in a test-tube; add to this a few crystals of ferrous
sulphate.   Shake the tube, incline it to one side, and cautiously pour a few drops of strong sulphuric acid down the
side of the tube, so that the heavy acid may sink to the
bottom, and there form a separate layer.   At the point of
contact of the two layers you will notice a dark brown-
coloured ring : this is proof of the presence of nitric acid ; it
is nothing else than the dark liquid you produced in the last
experiment by decanting nitrous fumes into a ferrous sulphate solution.   The reaction which goes on in the tube is
as follows :—The sulphuric acid acts on the nitre, nitric
acid and potassium sulphate are formed ; part of the ferr*ous*
sulphate acts on the nitric acid, robbing it of some of its

oxygen, fer*ric* sulphate and lower oxides of nitrogen are produced ; the latter dissolve in the excess of ferrous sulphate, and this solution, as you know, has a dark brownish-black colour. In applying the test, be careful to have all the solutions perfectly cold ; by pouring the acid gently down the side of the tube, you do not allow it to mix with the liquid, because if it did so, heat would be produced, and the dark colour destroyed, as you learned in Experiment II.

In this lesson you have learned—

(1) How to conduct a process of distillation.

(2) That by the action of sulphuric acid on potassium nitrate, nitric acid is produced.

(3) That this acid is a heavy, slightly coloured liquid, fuming in the air.

(4) What are the tests for nitric acid, and what is the mode of their application.

LESSON VI.

PREPARATION AND PROPERTIES OF NITROGEN MONOXIDE (NITROUS OXIDE).

Take the nitric acid remaining from last lesson, put it into a porcelain basin, dilute it with a little water, and add ammonia to it (stirring with a glass rod after each addition of ammonia), until a drop taken on the end of a rod just ceases to redden blue litmus paper. Set the basin on a piece of wire gauze resting on a tripod-stand over a flame, and evaporate down until the liquid becomes slightly viscid, and no longer exhibits signs of ebullition. During this evaporation add from time to time a few drops of ammonia, otherwise the solution becomes acid, and when you heat the ammonium nitrate formed you find red nitrous fumes produced. Set the basin aside—as the liquid cools it

becomes solid. When quite cold, break up the solid substance in the basin into small pieces.

In this process you have *neutralised* the nitric acid by
means of ammonia, the acid and alkali have combined together, the resulting *new* substance being ammonium nitrate
—the white salt in the basin. The substance produced by
such a union of an acid with a base is called a *salt*. As
this salt has a definite chemical composition, there must be
a point reached, in adding ammonia to the acid, at which
all the acid present is taken up by the ammonia ; this is the
point of neutralisation.

Put into a flask, which can be fitted with a delivery tube
(that used in the preparation of oxygen may be employed),
10 grams of the ammonium nitrate which you have just
made. Set the flask, fitted with its cork and tube, on the
retort-stand, and arrange the trough in the usual manner,
but use *warm water* instead of cold.

Heat the flask gently until the salt fuses, then gradually
increase the heat until the fused ammonium nitrate begins
to decompose with effervescence, and bubbles of gas succeed
each other rapidly through the water in the trough. Allow
a few minutes to elapse, then begin to collect the gas ; fill
three bottles, and set them aside on the small trays. In heating the flask, do not let the temperature get too high ; if you
see many white fumes appearing in the flask—especially
towards the close of the operation—the heat must be mode-
rated ; but do not withdraw the lamp entirely, or, as the flask
cools, the water may rush back from the trough into it, and
crack it.

By the action of heat ammonium nitrate is decomposed,
nitrogen monoxide is given off, and water remains ; you
may see this water condensing in the neck of the flask, and
trickling down the sides—$NH_4NO_3 = N_2O + 2H_2O$. This
nitrogen monoxide is slightly soluble in cold water, but much
less so in hot water, hence the reason for using warm water
in the trough.

Experiment I.—Lift one of the bottles from the tray, set it, mouth upwards, on the table, and plunge into it a lighted taper; the flame burns more brilliantly, but the gas itself does not take fire.   Do this experiment quickly, and immediately replace the bottle, inverted, on the tray.

Experiment II.—For this experiment you must pass the deflagrating spoon through a cork of a size to fit the mouth of the bottle ; below the cork on the spoon is fitted a small copper or tin plate, which prevents the burning phosphorus from setting fire to the cork.   (See fig. 14.)

Cut and dry a small piece of phosphorus, place it on the deflagrating spoon, set fire to it, and plunge it into the jar of gas used in the last experiment, fitting the cork on the spoon into the bottle. The phosphorus burns very brilliantly, with almost as bright a light as when it burned in oxygen.

Fig. 14.

This confirms the teaching of the last experiment. When the fumes produced in this experiment have somewhat subsided, plunge a lighted taper into the jar ; it is extinguished, the gas behaves therefore as nitrogen does : now pour a little blue litmus solution into the bottle, close the mouth with the palm of the hand, and shake well; the litmus is reddened, therefore an acid has been produced in this experiment.   This result agrees with that obtained by burning the phosphorus in pure oxygen (Lesson I.) or in air (Lesson IV.).   But there is a difference between the powers of supporting combustion possessed by this gas and oxygen.

Experiment III.—Place a small piece of sulphur on the spoon, set fire to it, but let it be only just lighted anb no more ; now bring it into the second bottle of nitrous oxibe : the flame is extinguished.   Had you employed oxygen in this experiment, you know that the sulphur would have burned with increased brilliancy.

These combustions in nitrous oxide are due to the fact that the body burning decomposes the gas, taking to itself the oxygen; but as there is a large quantity of nitrogen present, this dilutes the oxygen, and so the combustion is less violent than if it took place in oxygen alone. The oxygen and nitrogen being held together chemically, the low temperature of the burning sulphur was not sufficient to dissolve their union, the sulphur could get no oxygen, hence its flame was extinguished; but if you heat the sulphur considerably above its melting-point before plunging it into the gas, it will burn brilliantly.

A mixture of oxygen and nitrogen would diminish the intensity of combustion like the oxide of nitrogen; but to show the difference between such a mixture and nitrous oxide we will perform another experiment.

Experiment IV.—Pour a little cold water quickly into the third bottle of the gas, close the mouth with the hand, and agitate briskly; invert the bottle, still keeping the hand over the mouth, and bring it thus beneath cold water in a basin. Now withdraw the hand: you notice that the water rises in the bottle. thus showing that you have dissolved some of the gas in the water you added. Again slip the hand beneath the mouth of the bottle, lift it, containing water, out of the basin, agitate briskly and replace it in the basin as before; the water rises further in the bottle. Repeat this agitation until the water nearly fills the bottle. You now perceive how uneconomical it would have been to have collected the gas over *cold* water. This solubility in water also serves to distinguish the chemical compound of oxygen with nitrogen from a mechanical mixture of these gases which would not dissolve in water to the same extent.

You learn in this lesson—

(1) By what reaction nitrous oxide is produced.

(2) That this gas has the power of supporting combustion, and in what way it does this.

(3) That there is a difference between this gas
and oxygen in their respective powers of
supporting combustion.

(4) That there is a difference between a mechanical mixture and a chemical compound.

(5) That this gas is soluble in cold water.

(6) What is the meaning of the terms 'neutralisation' and 'salt.'

LESSON VII.

PREPARATION AND PROPERTIES OF NITROGEN DIOXIDE (NITRIC OXIDE).

INTRODUCE about 15 grams of copper in small pieces into a
flask which is fitted with a delivery and funnel tube in the
manner represented in fig. 4.   Mix 30 c.c. of water with the
same amount of strong nitric acid, and pour about three-
fourths of this mixture on to the copper clippings through
the funnel tube.   Action evidently goes forward in the flask,
for it is almost immediately filled with red vapours, and bubbles of gas make their way through the water in the trough;
the red vapours in the flask soon however almost entirely
disappear.   Now begin to collect the gas which is coming
over, and fill four bottles with it.   This gas is produced by
the above action without heating the flask.   Should the
action become too violent, pour a little cold water on the
outside of the flask; if the stream of gas slackens, add a
little more of the dilute nitric acid.

Experiment I.—Plunge a lighted taper into one of the
bottles; the taper goes out, and the gas does not take fire: you
notice, however, that ruddy fumes appear in the bottle.   The
formation of these will be explained by the fourth experiment.

Experiment II.—Get ready a small piece of phosphorus
on the spoon, light it, and plunge it, *immediately* it begins to

burn, into the second bottle of the gas ; the phosphorus is extinguished.   Nitric oxide does not support combustion as nitrous oxide does.   If you repeat this experiment, allowing the phosphorus to burn briskly for a little time before you bring it into the gas, holding it for a few moments in the Bunsen flame, it will burn with increased brilliancy in the nitric oxide.   You thus see that the temperature at which nitric oxide is decomposed and made to part with its oxygen is considerably higher than that at which nitrous oxide is similarly split up ; the latter therefore much more readily supports combustion than the former gas.

Experiment III.—Into the second bottle of nitric oxide plunge a piece of brightly-burning sulphur, held in the deflagrating spoon : it is extinguished.   In nitrous oxide under the same circumstances the sulphur continues to burn.

Experiment IV.—For this experiment you will require a small bottle of oxygen (make this as directed in Lesson I., using a smaller flask, and about 10 grams of potassium chlorate with 2 grams of manganese dioxide).   Decant about one-fourth of the nitric oxide contained in the third bottle into another gas bottle, by means of the trough, and very slowly pass up into this a little oxygen.   As each bubble of oxygen comes into contact with the nitric oxide, brown ruddy fumes are produced.   After the addition of a small quantity of oxygen, shake the bottle containing the mixed gases (taking care not to let its mouth get above the water, else air will rush in) ; the water rises in the bottle, showing that gas is being dissolved.   Continue the addition of oxygen, and at intervals shake the bottle ; you find that at last it is entirely filled by the water.   You would naturally expect that by the addition of one gas to another, the volume of gas would be increased ; in this case, however, it is diminished until it is actually *nil*.   The reason is that the oxygen combines with the nitric oxide, forming nitrogen tetroxide, NO_2, a gas of a brown colour—hence the brown fumes in the flask when you prepared nitric oxide, and in the bottle of this gas when you exposed it to the air in bringing the taper into it. This gas is very soluble in water, so that as fast as it was produced in the last experiment, it was dissolved. If then you add this gas nitric oxide to a measured quantity of a mixture of oxygen and nitrogen, these ruddy fumes will be produced ; if you now shake up with water, the fumes will be dissolved, and a contraction in the bulk of the gases will ensue ; by measuring this contraction you may determine how much oxygen was present in the original mixture.

On this principle analyses of air were formerly made. To illustrate the method of carrying out such analyses we will try—

Experiment V.—Take a glass tube about 30 cm. long and 2 cm. in diameter, divide it into five equal parts, each part corresponding to the contents of a smaller tube 1 cm. diameter and 14 cm. long ; mark the divisions on the outside of the large tube by slipping over it small india-rubber rings. Pour water into the large tube, so that when inverted over the trough the water shall reach to the first ring (see fig. 15); set the tube on the beehive shelf, and fill the smaller tube, over the trough, with nitric oxide ; pass this into the larger tube, again fill the smaller tube with nitric oxide, and again pass this into the larger, in which you now have five volumes of air and two volumes of nitric oxide ; shake this tube briskly; the oxygen of the air forms with the nitric oxide, nitrogen tetroxide, which is dissolved in the water.

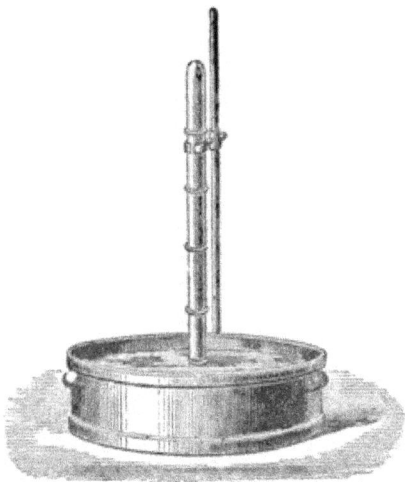

FIG. 15.

Add a few more bubbles of nitric oxide ; if no red fumes

are formed, the reaction is ended; again shake the tube
and depress it so that the water inside and outside the
tube may be at the same level; you find that the air or gas
in the tube now occupies only four volumes—one volume
of oxygen has therefore been taken away.   Hence you
conclude that five volumes of air contain one volume of
oxygen.

Pour the liquid remaining in the flask used in preparing
the nitric oxide into a porcelain basin, evaporate it nearly
to dryness in a draught chamber, and set it aside.   After
some time you notice a blue solid substance formed in the
basin—this is copper nitrate, produced by the action of the
nitric acid on the copper: it was kept in solution by the
excess of dilute nitric acid which you have evaporated off.

From these experiments you learn—

(1) What is the method of preparing nitric oxide.

(2) What is the behaviour of this gas towards
combustible bodies.

(3) That a high temperature is required to decompose this gas.

(4) That with oxygen, nitric oxide forms a brown-
coloured   gas—nitrogen   tetroxide—very
soluble in water.

(5) That the formation of this nitrogen tetroxide
by adding nitrogen dioxide to oxygen in a
systematic manner may be made available
for the analysis of air.

LESSON VIII.

PREPARATION AND PROPERTIES OF AMMONIA.

In this lesson you have to deal with a gas which is so
soluble in water that it cannot be collected over the trough.
It differs much in weight from air; we can therefore use, so

to speak, an air trough in collecting the gas. This trough is the atmosphere around us; a gas bottle is already filled and standing in this trough; but as the gas to be collected is lighter than air, the bottle must be inverted, so that when the gas passes upwards into it the heavier air may be driven out at the mouth which opens downwards.

Mix a few grams of ammonium chloride with about an equal amount of lime, heat this mixture gently in a dry test-tube; you recognise the well-known smell of ammonia. Repeat this experiment, using a little caustic soda instead of lime; the same effect is produced. Moisten a small piece of red litmus paper (reddened by holding it for a moment in the fumes coming from a bottle of nitric acid), and hold it over the test-tube, it is changed to blue again; warm it gently, the blue colour disappears.

Weigh out 10 grams of ammonium chloride (sal ammoniac), and powder it in a mortar; also weigh out 20 grams of powdered quicklime.

Choose a flask of the same shape as that used in Lesson I. and a cork to fit it, and bore in the latter a neat round hole; take a piece of glass tubing about 30 cm. long, bend it at right angles, the length of the smaller limb being 5 cm.; at a distance of 10 cm. from this bend make another in the same direction, and let the third limb of the tube be 15 cm. long. The smaller limb of this tube is fitted into the cork of the flask.

Into a small wide-mouthed bottle fit a good cork having two holes bored in it, through one of which passes a piece of tolerably wide tubing (4 or 5 cm. long), within which the longer limb of the first tube passes nearly to the bottom of the small bottle. The shorter limb of a third tube, similar to the first, except that the bends are in opposite directions, is fitted into the second hole in the cork of the small bottle, into which you now pour about 20 c.c. of ordinary ammonia solution.

Mix the sal ammoniac with two-thirds of the total quantity of the lime in a mortar, transfer the mixture to the

flask, and then add the remainder of the lime; fit the cork
into the flask, and put the pieces of apparatus together, the
flask standing over wire gauze on a tripod or retort-stand,
while the small bottle is supported on a block of wood.   A
perfectly dry gas bottle is supported, mouth downwards,
on another retort-stand, so that the tube leading from the
small bottle may pass nearly to the top of it.   The apparatus
is shown in fig. 16.

By the application of a gentle heat to the flask, the sal

F IG. 16.

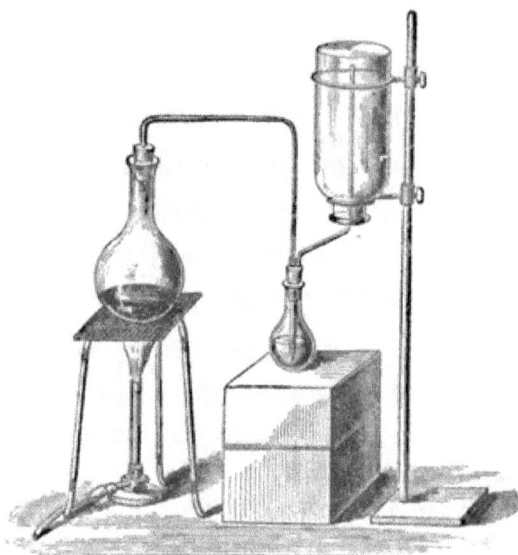

ammoniac and quick lime are caused to react on one
another, with the production (1) of ammonia—the first
portions of which are dissolved by the water in the small
bottle (this at first not being quite saturated), and which is
afterwards collected in the gas bottle ; (2) of water—which
is absorbed by the upper layer of quick lime ; and (3) of
calcium chloride—a white solid which remains in the flask
mixed with a little undecomposed quick lime and sal ammoniac, $CaO + 2NH_4Cl = CaCl_2 + 2NH_3 + H_2O.$

The liquid in the small bottle serves to wash the gas. The ammonia passes upwards into the gas bottle, and gathering near the top, gradually drives out the air, until at last the bottle is filled with ammonia ; when this is done, the gas will of course flow over the edges of the mouth of the bottle. You may ascertain that this is taking place by holding a piece of moistened turmeric paper close to the outside of the bottle, a little above the mouth ; it will be turned brown, showing the presence of ammonia. Now remove the bottle from the retort-stand, and set it mouth downwards on a glass plate. Fill in this way three bottles with the gas.

Experiment I.—Pass a lighted taper into one of the bottles of ammonia, supported, mouth downwards, on a retort-stand ; you notice a momentary flash of greenish-coloured flame round the taper, which is then immediately extinguished. Ammonia tends to burn when a flame is brought near it.

By passing ammonia mixed with oxygen through a glass tube, and applying a light to the issuing gas, it will burn briskly.

Experiment II.—Choose a piece of tubing about 100 cm. long and $1\frac{1}{2}$ cm. in diameter ; support this by means of a clamp, so that the end of the delivery tube from a flask containing a little strong ammonia solution may pass a little way into it, and let it be slightly inclined at an angle with the surface of the table. Fit up the oxygen apparatus used in Lesson I. (using smaller quantities of potassium chlorate and manganese), and let the end of this delivery tube also pass into the wide tube—see fig. 17. Heat the two flasks so as to produce a gentle stream of oxygen, and also of ammonia ; you will soon have a mixture of these two gases issuing from the upper end of the wide tube. On applying a light to the opening, the ammonia burns with a long greenish flame.

Experiment III.—Pour 30 or 40 c.c. of strong ammonia solution into a flat-bottomed flask capable of holding half a

litre ; wind a piece of thick platinum wire, about 30 c.m.
long, round a glass tube so as to form a spiral, and arrange
this on a glass rod so that it shall hang into the flask to within
a centimetre or so of the ammonia solution.   Make the end
of this spiral red-hot in the flame of a Bunsen lamp, and then
plunge it into the flask.   You see that it continues to glow,
while the flask gradually becomes filled with white fumes.

In the last experiment you caused ammonia to combine
rapidly with oxygen ; in this a similar combination takes

FIG. 17.

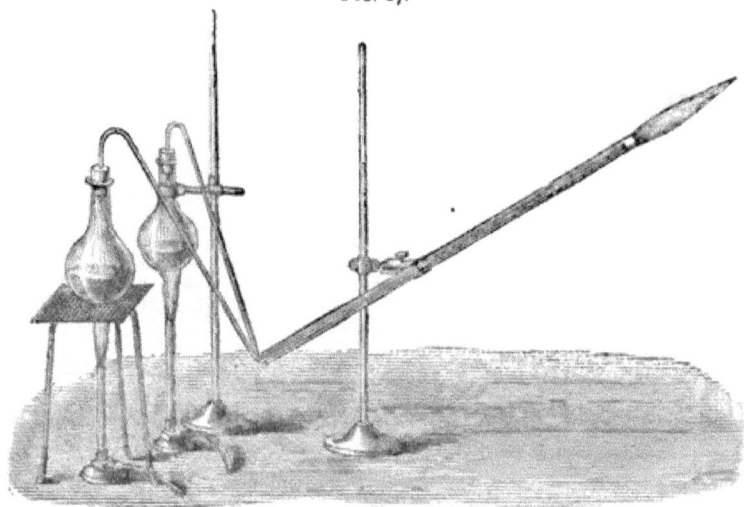

place, but much more slowly.   Under the influence of the
heated platinum, the oxygen of the air combines with the
ammonia, forming *nitrous acid*, which in turn reacts on the
excess of ammonia and combines with it to form the salt
ammonium nitrite, which you see appearing as white fumes
in the flask.

The next experiment will show you the formation, in a
somewhat similar manner, of another ammonium salt.

Experiment IV.—Take 3 or 4 c.c. of strong hydrochloric
acid, pour it into a small flask, set this on wire gauze on
a tripod, and heat the solution.   While the fumes of hydro-

chloric acid are coming off, bring the second bottle of ammonia, mouth downwards, over the flask; it is immediately filled with dense white fumes. The hydrochloric acid combines with the ammonia, forming a new substance, ammonium chloride ; such a union of an acid with a base is, as you already know, called a salt.

Experiment V.—Lift the remaining bottle of ammonia, mouth downwards, standing on a glass plate, into the trough filled with water, withdraw the plate and shake the bottle ; the water will quickly rise until it nearly fills the jar, showing the great solubility of ammonia in water.

In this lesson you have learned—

(1) That gases soluble in water may be collected by displacement, and how to collect such gases.

(2) By what reaction ammonia is produced.

(3) What is the action of this gas on vegetable colouring matters.

(4) That ammonia is a combustible gas, especially when supplied with a large amount of oxygen; that it also may be made to undergo slow combustion.

(5) That ammonia is very soluble in water.

(6) That ammonia gas is much lighter than air.

(7) That ammonia readily combines with certain acids, producing thereby salts of these acids.

LESSON IX.

PREPARATION AND PROPERTIES OF CARBON DIOXIDE.

Fit a flat-bottomed flask with a cork, through which pass a funnel tube and a delivery tube, this latter being bent twice at right angles, so that the shorter limb may be about 8 cm., and the longer 20 cm. in length. Weigh out 30 grams of marble; break it up into pieces the size of a pea, and

FIG. 18.

place these in the flask (remembering the precautions given in Lesson II.), fit the cork with the tubes in its place, and set the flask so that the longer limb of the delivery tube may dip nearly to the bottom of a gas bottle standing on a block of wood, and having its mouth covered with a piece of pasteboard through a hole in which the tube passes (fig. 18). Pour a little water into the flask, and then some strong hydrochloric acid; a brisk effervescence ensues, a gas being evolved. If after the expiry of a few moments you plunge a lighted taper into the gas bottle, you will find that as it approaches the bottom of the bottle it goes out ; there is evidently some gas collecting here, and as it gathers at the bottom of the bottle, it must be a heavy gas. The atomic weight of carbon dioxide, CO_2, is 44 ; it is therefore $\frac{44}{2} = 22$ times heavier than hydrogen. But hydrogen is 14·47 times lighter than air, carbon dioxide is therefore about $1\frac{1}{2}$ times heavier than air.

Continue the evolution of gas for some time, and again bring the lighted taper into the jar; as soon as the flame comes within the jar it is extinguished ; you have thus filled the bottle with a gas heavier than air by *downward displacement.* Fill three bottles and a small flask capable of holding about 100 c.c. with the carbon dioxide, cover them with well-greased glass plates, and set them aside.

Experiment I.—Pour a little clear lime-water into a small beaker, dip the end of the delivery tube into this liquid, and allow the carbon dioxide to bubble through it; you very soon perceive that the clear solution becomes turbid—the reason for this has been explained in Lesson I. Continue the passage of the gas, the turbidity after a time disappears. When this occurs, withdraw the delivery tube, and boil the liquid in the beaker; the turbidity soon again makes its appearance. The disappearance of the turbidity on the continued passage of the gas is owing to the solution of the calcium carbonate in the carbonic acid ; on boiling, the solution is decomposed, carbon dioxide is evolved, and the insoluble calcium carbonate is reprecipitated.

Experiment II.—Pour a little blue litmus solution into a small flask, and pass the carbon dioxide through this liquid. You notice that it becomes a wine-red colour, differing entirely from the pure red produced by the action of nitric or hydrochloric acid upon litmus solution. Boil the reddened liquid, it becomes blue again. Now set aside, for further use, the flask employed in the preparation of the gas.

Experiment III.—Place on the table a little piece of lighted taper stuck into a cork, withdraw the glass plate from the mouth of one of the bottles of carbon dioxide, and gradually invert the bottle as if you were pouring from it on to the taper (fig. 19). The lighted taper soon goes out.

FIG. 19.

This experiment confirms what you have already learned while preparing this gas, viz. that it extinguishes flame, and that it is heavier than air.

Experiment IV.—Put a small newly-cut piece of sodium, about the size of a pea, into the flask containing carbon dioxide, and gently heat the flask—the sodium melts, and then takes fire ; cover the mouth of the flask with the thumb, and agitate it, so as to move the burning sodium about from place to place. When the sodium has ceased to burn, allow the flask to cool, and then pour a little water into it ; you notice black soot-like flakes floating about in the water. The carbon dioxide has been broken up into its elements by the burning sodium, which seized upon the oxygen so greedily as to cause evolution of light and heat, while the carbon appeared in the free state. Carbon dioxide, in respect of its power of supporting combustion, may be classed with nitric oxide.

Experiment V.—Pour a little water into the second bottle of carbon dioxide, close the mouth of it with the hand, and shake briskly ; bring the bottle, inverted, under the water of the trough, and withdraw the hand : the water rises in the jar. Now place the hand as before, and lift the bottle out of the water ; pour into it a few drops of clear lime-water, and shake several times ; you notice that the liquid becomes turbid, which, as you know, is proof of the presence of carbon dioxide. You have therefore dissolved this gas in water.

Experiment VI.—Into the third bottle of carbon dioxide pour a little caustic potash or caustic soda solution, close the mouth of the bottle with the hand, and shake briskly ;

bring now the bottle, inverted, under the water in the trough, and withdraw the hand. By the fact that the water rushes into the jar, you learn that the gas has been absorbed by the caustic potash.

Pour the liquid in the generating flask into a porcelain basin, and evaporate it over a flame to complete dryness, stirring the mass from time to time with a glass rod, so as to have all parts equally heated ; do not allow it to gather into a hard cake, else you will have difficulty in getting it out of the basin. Put this solid substance, when quite dry, into a wide-mouthed stoppered bottle, labelling it 'calcium chloride.' This calcium chloride is the second product of the action of hydrochloric acid on marble (calcium carbonate) :

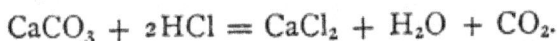

$$CaCO_3 + 2HCl = CaCl_2 + H_2O + CO_2.$$

You have learned then—

(1) How to collect a heavy gas by *downward displacement.*

(2) That marble, when acted upon by an acid, gives off the gas carbon dioxide, CO_2.

(3) That this carbon dioxide is a heavy gas.

(4) That it extinguishes flame.

(5) But that certain substances can, at high temperatures, decompose this gas, taking its oxygen to themselves.

(6) What is the action of carbon dioxide on lime-water and on litmus solution.

(7) That carbon dioxide is soluble in water, and how to recognise it in this solution.

(8) That this gas is absorbed by caustic potash or soda solution.

LESSON X.

PREPARATION AND PROPERTIES OF CARBON MONOXIDE.

PUT a few crystals of oxalic acid into a dry test-tube, drench them with strong sulphuric acid, and heat gently over a flame : you notice an effervescence in the tube ; continue heating for a few moments, then bring a lighted taper to the mouth of the tube : a pale lavender-blue flame appears for an instant, passes down the tube, and goes out. Pour a little clear lime-water into another test-tube, and decant the gas (see Lesson V.) from the first tube (heating this all the while) into the lime-water; this soon becomes turbid : carbon dioxide is therefore present. But carbon dioxide is not an inflammable gas ; we have therefore two gases produced in this reaction : one of them is carbon dioxide, the other carbon monoxide. Arrange the apparatus in which you prepared hydrogen (Lesson II.), setting the flask on a retort-stand. Place 10 grams of crystallised oxalic acid in the flask, and add, through the funnel tube, about 15 grams of strong sulphuric acid ; heat the flask gently, and after the gas has been coming off for a few minutes, begin to collect it over the trough. Fill a bottle with the mixed gases, and set it aside, mouth downwards. on a small tray. (Set the generating flask *at once* in the draught chamber, carbon mon-oxide being a very poisonous gas.)

Experiment I.—Pour about 20 c.c. of a strong solution of caustic potash rapidly into the bottle, close its mouth with the hand, and shake briskly several times. You feel by the pressure of the outside atmosphere on the hand that some gas has been absorbed by the potash within the bottle. You have already seen that carbon dioxide is produced by the action of strong sulphuric acid upon oxalic acid ; the caustic potash absorbs all this carbon dioxide, leaving the monoxide untouched. Now invert the bottle beneath the water in the trough, and withdraw the hand; the water rushes in until it fills one half of the bottle.

Sulphuric acid heated with oxalic acid gives rise therefore to equal volumes of carbon monoxide and dioxide. The sulphuric acid acts by taking away the elements of water, $C_2H_2O_4 - H_2O = CO_2 + CO$.

Experiment II.— Decant the gas remaining in the bottle used in the last experiment into another smaller bottle, and bring a light to its mouth; the gas burns with the peculiar pale blue flame noticed in the preliminary experiments. When the flame has gone out, add a little lime-water to the bottle, and shake it briskly. A turbidity tells you that carbon dioxide is present. By burning carbon monoxide, therefore, carbon dioxide is produced. This is in keeping with what you have already learned about combustion. The monoxide, CO, has been burned or oxidised, a compound relatively richer in oxygen being produced.

There is another method of producing this carbon monoxide which is one of much interest. Take about 10 grams of formic acid, and add to this, in the apparatus you have just used, 15 grams of strong sulphuric acid. Gently heat the flask, and collect over the trough a bottle full of the gas which is evolved. Test this gas, first by shaking it with caustic potash—nothing is absorbed; then apply a light, when you at once see, by the colour of the flame of the burning gas, that it is carbon monoxide. By the action of sulphuric acid upon formic acid, pure carbon monoxide is produced. Formic acid, H_2CO_2, gives up to sulphuric acid the elements of water, H_2O, while carbon monoxide, CO, remains.

$$C(OH)_2 = CO + H_2O.$$

Formic acid may be prepared by heating together for several hours glycerine, oxalic acid, and a little water, then adding more water, and raising the temperature : liquid formic acid distils over. The same amount of glycerine may be again used to effect the transformation of a fresh quantity of oxalic acid. This formic acid was formerly obtained from ants, or from certain plants, *e.g.* nettles—in which it exists; sometimes also by the reaction mentioned above ; but in this case the oxalic acid was itself obtained from plants, so that formic acid was entirely of animal or vegetable origin. But the experiment you have just performed, viz., splitting this acid into carbon monoxide and water, suggested the idea that by combining these two substances together, formic acid might be re-formed. This has actually been done, and the chemist has thus built up from *inorganic* or *mineral* constituents a substance formerly obtainable only from organised forms. Such a building up is termed *synthesis,* and is exactly opposed to *analysis,* or pulling to pieces.

You may have noticed on a winter's evening a pale blue flame playing on the surface of a coal fire, when the coals are red-hot, and little or no smoke is emitted from them. This flame is produced by burning carbon monoxide. You have already learned (in Lesson I.) that by burning carbon, carbon dioxide is produced ; by burning coals, therefore, this gas is formed; but this carbon dioxide in passing through the layer of red-hot coals in the upper part of the fire is robbed of part of its oxygen, carbon monoxide being produced, while the oxygen unites with another atom of carbon to form also this gas carbon monoxide ; thus the ultimate form assumed by the gas produced in the combustion of coal in such a fire as that described is that of carbon monoxide :

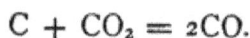

$$C + CO_2 = 2CO.$$

From the experiments in this lesson you learn—

(1) What is the action of sulphuric acid upon oxalic acid.

(2) That carbon monoxide is a combustible gas, producing, when burned, carbon dioxide.

(3) That carbon dioxide can be separated from the monoxide by shaking the mixed gases with caustic potash solution.

(4) That carbon monoxide is produced in a coal fire.

(5) What is the action of sulphuric acid on formic acid.

(6) What is the meaning of the term synthesis.

LESSON XI.

PREPARATION AND PROPERTIES OF CHLORINE.

HEAT a few grams of manganese dioxide with a little hydrochloric acid in a test-tube; a yellowish green gas * with an extremely pungent odour is evolved:

$$MnO_2 + 4HCl = MnCl_2 + 2H_2O + Cl_2.$$

A more uniform stream of chlorine gas is obtained by acting on a mixture of manganese dioxide and sodium chloride with dilute sulphuric acid:

$$2NaCl + MnO_2 + 2H_2SO_4 = Na_2SO_4 + MnSO_4 + 2H_2O + Cl_2.$$

Weigh out about 30 grams of manganese dioxide and the same amount of sodium chloride, mix them in a mortar, and transfer the mixture to a large flat-bottomed flask. Mix cautiously 38 grams of water with 60 grams of strong sulphuric acid in a large beaker, and allow this mixture to cool. The apparatus is similar to that in which you prepared carbon dioxide (Lesson IX.), but the generating flask must be considerably larger. You know that chlorine has the atomic weight 35·5; it is therefore $35\frac{1}{2}$ times heavier than hydrogen, but hydrogen is $14\frac{1}{2}$ times lighter than air. How therefore must chlorine be collected: by upward or downward displacement? When you have settled the arrangement of the receiving bottles, pour the cold mixture of

* Hence the name, from χλωρὸς = yellowish green.

sulphuric acid and water through the funnel tube into the
flask ; an action at once ensues, the flask gradually becomes
filled with yellowish-green vapours, which pass over into the
receiving bottle.   Conduct all operations with chlorine in a
draught chamber, as the fumes of this gas are very hurtful.
Make sure that all the connections in the apparatus are
tight. (Should you get a whiff of the gas into the lungs,
pour a little alcohol on to a piece of filtering paper, and hold
this over the mouth and nose so as to inhale the vapours.)
When, judging by the colour, you think that the bottle is
full of chlorine, remove it, cover it with *a well greased* glass
plate, and set another in its place.   Fill thus seven bottles,
and proceed to test the properties of the gas.

Should the flow of gas slacken, apply a gentle heat to
the generating flask by means of a small Bunsen lamp ; the
action will soon go on briskly again.

Experiment I.—Invert the first bottle of chlorine in the
trough, withdraw the plate, and shake the bottle ; the water
gradually rises in the jar, showing you that this gas is very
soluble in water.   It is this property of chlorine that obliges
us to have recourse to collection by displacement.

Experiment II.—Moisten a piece of madder-dyed red
cloth with water, and put it into one of the bottles of
chlorine ; cover the mouth of the bottle with a glass plate,
well greased where it touches the bottle, and notice how the
red colour of the cloth gradually disappears ; allow the
bleaching action to go on, and proceed with—

Experiment III.—Quickly pour a little strong sulphuric
acid into the third bottle of chlorine, replace the glass plate,
and shake the bottle several times ; after a little time introduce into the bottle a piece of dyed cloth, similar to that
used in last experiment.   It is not bleached.   The sulphuric
acid, by taking away all moisture present in the jar, dries
the chlorine, and this gas, when perfectly dry, does not
bleach.

Experiment IV.—Moisten a strip of filtering paper with

turpentine, and plunge it into the third bottle of chlorine ; a cloud of black smoke is at once produced, accompanied by a momentary flame, and you find the paper charred and blackened in the jar. The affinity which chlorine possesses for hydrogen is very great, and as turpentine consists of carbon and hydrogen chemically combined, the hydrogen is seized upon by the chlorine, and separated from the carbon or charcoal, which, as you know, is a black soot-like substance. With another bottle of chlorine you may perform an experiment showing the great affinity of chlorine for metals.

Experiment V.—Place a little powdered metallic antimony on a small piece of paper, and from this shake it slowly into the bottle containing chlorine. The antimony, as it falls, burns, each little grain sparkling brilliantly. Chlorine and antimony have therefore very great affinity for each other, so great that light is produced by the intensity of their combination.

Experiment VI.—Into another bottle of chlorine plunge a piece of phosphorus supported on the deflagrating spoon ; the phosphorus takes fire and burns with a brilliant flame, showing you that chlorine has a great affinity for this substance as well as for so many others.

Experiment VII.—Put two pieces of paper into the last bottle of chlorine with some letters printed or written on each, those on one paper being printed with common ink, those on the other with printers' ink.

You notice that the common ink is soon bleached and almost entirely disappears, while the letters formed with printers' ink remain unacted upon. This is a further illustration of the affinity of chlorine for hydrogen. Common ink is a vegetable substance, and contains a considerable quantity of hydrogen united with other elements ; while printers' ink is essentially a mixture of carbon, in the form of lampblack, with some thickening material.

Experiment VIII.—Place a few grains of powdered iron

pyrites in a test-tube with a little water, and, while you shake
the tube so as to keep the solid particles suspended in the
water, pass a gentle stream of chlorine through the liquid.
(The materials used in preparing chlorine for the foregoing
experiments will probably, if heated, yield enough of the gas
for this experiment.) The pyrites rapidly disappears, and
you have soon only a few fine white particles floating about ;
allow these to settle, pour the clear liquid into another test-
tube, divide it into two portions, and apply a chemical test
to each.

To portion (1) add a little barium chloride ; a dense
white precipitate immediately forms. This, as you will hereafter more fully learn, is proof of the presence of *sulphuric
acid*.

To portion (2) add a little ammonia ; a foxy-red precipitate tells you that iron, in a high state of oxidation, is
present.

Iron pyrites is essentially sulphide of iron. The action
of the chlorine has been to combine with the hydrogen of
the water, while the oxygen thus liberated has attacked the
iron and sulphur and oxidised them.

You thus see how chlorine may be made an exceedingly
valuable *oxidising* agent.

From these experiments then you learn—

(1) How to prepare chlorine.

(2) That this gas is very soluble in water.

(3) That chlorine has a great affinity for many substances, especially for hydrogen.

(4) That this property may be applied to bleaching
purposes.

(5) That chlorine may be made to act as an oxidising agent.

LESSON XII.

PREPARATION AND PROPERTIES OF HYDROCHLORIC ACID.

PLACE a few grains of common salt in a test-tube, and add a drop or two of strong sulphuric acid; a brisk effervescence ensues, and a gas having a strongly acid pungent odour is given off. The semi-fluid mass in the test-tube solidifies on cooling, forming hydrogen-sodium-sulphate. This reaction is exactly analogous to that by which you prepared nitric acid, NO_3 being in this instance replaced by Cl, and K by Na:

$$NaCl + H_2SO_4 = NaHSO_4 + HCl.$$

Into the flask used in preparing carbon dioxide (see Lesson IX.) put about 20 grams of common salt in lumps. To get the salt into this state, fuse it in an iron ladle in the furnace, and when cool break it up into pieces. Pour, through the funnel tube, strong sulphuric acid sufficient to cover the salt in the flask, which is set on a retort-stand, and place a bottle to collect the gas, as in the case of carbon dioxide, testing when the bottle is full by bringing a lighted taper near the mouth of the bottle; if the light is extinguished, remove the bottle, and cover it with a greased glass plate. Fill thus two bottles with the gas. If the action slackens, apply a gentle heat.

You have already noticed, while testing if the bottles were full of gas, that hydrochloric acid is incombustible, and that it does not support combustion.

Experiment I.—Bring a piece of blue litmus paper into the first bottle of the gas, the paper is at once reddened; this property of reddening litmus you already know to be characteristic of acids.

Experiment II.—Bring the other bottle, inverted, beneath the water in the trough, withdraw the plate and shake the bottle briskly; it is rapidly filled by the water rising within it. This experiment shows you the great solubility of hydrochloric acid in water. Ordinary hydrochloric acid consists of a solution of the gas in water. That the gas is to a certain extent expelled from such a solution, on boiling, you have already had proof.

Experiment III.—To prepare such a solution, fit up an apparatus similar to that used in making the ammonia solution, but in the small bottle put a little ordinary dilute hydrochloric acid, and let the delivery tube from this bottle dip downwards into a receiving bottle, which is partially filled with water.

The fumes formed when hydrochloric acid gas issues into the air are caused by the condensation of the gas by the moisture in the atmosphere.

The neutralisation of this acid by ammonia and consequent production of the salt ammonium chloride, you have also already noticed.

Experiment IV.—Fill a bottle with hydrogen, and another over the trough, using *warm* water, with chlorine; pass equal volumes of each of these into a soda-water bottle, cork it, surround it with a cloth, and shake it well ; withdraw the cork and bring a light to the mouth of the bottle ; a loud explosion ensues, and the formation of hydrochloric acid is rendered evident by the white fumes which appear on bringing a little ammonia to the mouth of the soda-water bottle. Hydrochloric acid gas is made up therefore of equal volumes of hydrogen and chlorine.

Experiment V.—Heat, in a test-tube, a little strong hydrochloric acid with a few crystals of potassium dichromate ; by the colour and smell of the issuing gas you at once recognise it to be chlorine. As in Experiment IV. you built up hydrochloric acid from hydrogen and chlorine, so in this experiment you have broken up the acid, liberating its constituent elements :

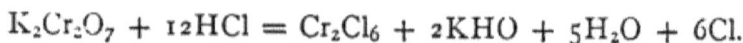

$$K_2Cr_2O_7 + 12HCl = Cr_2Cl_6 + 2KHO + 5H_2O + 6Cl.$$

In this lesson you learn—

(1) By what methods hydrochloric acid can be prepared.

(2) That this gas is soluble in water, and the practical application of this fact.

(3) What is the action of hydrochloric acid upon litmus.

(4) That fumes are caused when hydrochloric acid gas issues into the air, and the reason of this.

(5) What is the synthetical method of preparing this gas, and what we thus learn of its formation.

(6) What is the analytical method of breaking up hydrochloric acid.

LESSON XIII.

PREPARATION AND PROPERTIES OF OXYGEN DERIVATIVES OF CHLORINE.

FIT up an apparatus for generating chlorine, using half as much salt, manganese, &c., as directed in Lesson XI. Make about 200 cc., of a tolerably strong solution of caustic potash, place this in a beaker which is kept cool by being surrounded with water, and lead the chlorine gas which is generated into the liquid. After a few minutes remove the beaker, pour about half of its contents into another dish, place the beaker, resting on wire gauze, on a stand with a lamp beneath, and continue to pass in the chlorine gas, keeping the solution boiling. Meanwhile make a few experiments with that part of the solution which you have set aside.

First notice its smell: it is sensibly different from that of chlorine, although chlorous.

Experiment I.—Dip a piece of madder-dyed cloth into the solution : it is not bleached.

Experiment II.—To a small quantity of the solution in a test-tube add a drop or two of hydrochloric acid. Action ensues, and a gas is given off which, by its smell and colour, you at once recognise to be chlorine.

Experiment III.—Again dip the piece of cloth used in Experiment I. into the solution, then pass it quickly through dilute hydrochloric acid, and afterwards wash it in water; it is now bleached. You see that, by the action of the acid on the solution contained in the pores of the cloth, chlorine is liberated; as the cloth is moist, it is bleached by this chlorine.

The solution you have just been experimenting with is a mixture of potassium hypochlorite and chloride: $2KHO + Cl_2 = KClO + KCl + H_2O$. Bleaching powder is the corresponding calcium compound. To illustrate its formation, perform—

Experiment IV.—Fill a dry flask with chlorine by downward displacement, put into it a little powdered quicklime, and shake the flask; the yellow gas gradually disappears, being absorbed by the lime (as it was by the caustic potash). Pour a little water into the flask, and shake it up with the bleaching powder which you have made; then test the bleaching effect of the solution, as in Experiment III.

You now understand the object of dipping the goods into acid (*souring*, as it is termed) after passing them through the bleaching powder solution.

To the remainder of the solution of potassium hypochlorite add a little very dilute nitric acid, place the mixture in a retort to which a small receiver is adapted, and apply a gentle heat. After a little time a colourless peculiarly smelling liquid condenses in the receiver. This is hypochlorous acid, $HClO$.

Experiment V.—Dip a piece of dyed cloth into the distillate you have just obtained; the colour of the cloth is at once discharged.

Experiment VI.—Add a few drops of dilute hydrochloric

acid to another portion of the liquid ; chlorine is generated. You can now more fully understand the action of bleaching powder. When the goods are soured, the first effect of the acid probably is to produce hypochlorous acid—just as you have done ; but this (as you have shown in Experiment V.) is at once decomposed, chlorine being liberated.

You may now put together all you have learned about chlorine, and attempt to answer the question, How does chlorine bleach? If you consider that dry chlorine gas alone does not discharge vegetable colouring matters (Lesson XI., Experiment III.) ; that chlorine has an intense affinity for hydrogen (Lesson XI., Experiment IV.) ; that chlorine in presence of moisture does bleach (Lesson XI., Experiment II.) ; and that moisture or water consists of hydrogen and oxygen (Lesson III., Experiment I.)—you will be led to the conclusion that the bleaching action of chlorine must depend upon its power of combining with the hydrogen of the water present, and thus liberating oxygen. At the moment of its liberation oxygen is possessed of peculiarly active properties (it is in this state called *nascent* oxygen), so that it combines readily with the colouring matter of the cloth, oxidising this to form compounds which are removed in the subsequent washing.

Return to the heated potash solution through which you have been passing chlorine ; it will now be saturated with the gas. Remove the chlorine-generating apparatus, pour the solution into a basin, evaporate it to about half its bulk, and set it aside. A salt crystallises out on cooling ; drain off the liquid, and dry the crystals between folds of filtering paper. These are crystals of potassium chlorate,

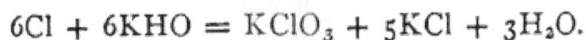

$$6Cl + 6KHO = KClO_3 + 5KCl + 3H_2O.$$

The potassium chloride formed, being more soluble than the chlorate, remains in solution. You have thus separated two salts by taking advantage of their different degrees of solubility in water.

Experiment VII.—Take one or two of the crystals you
have prepared, place them in a test glass, cover them

FIG. 20.

with water, put into the glass a very small
piece of phosphorus, and set a funnel tube in
the glass so that its lower extremity just touches
the crystals (see fig. 20), pour a few drops of
strong sulphuric acid down the funnel tube ; a
greenish gas is evolved which, coming in contact with the phosphorus, attacks it so violently
as to cause it to take fire even under water.
This gas is another oxide of chlorine, viz. the
tetroxide, Cl_2O_4. You have here an instance of
very intense chemical action.

In this lesson you learn :—

(1) The modes of producing some of the oxygen
derivatives of chlorine, and their leading properties.

(2) One of the methods for separating two salts
occurring together in solution.

(3) The *modus operandi* of bleaching, and the way
in which chlorine bleaches.

LESSON XIV.

PREPARATION AND PROPERTIES OF BROMINE.

INTO a small stoppered retort pour a mixture of 3 grams of
manganese dioxide with $1\frac{1}{2}$ grams of potassium bromide, set
the retort on wire gauze on the ring of a stand, allowing its
beak to pass into a small receiving flask, which is kept cool
by being surrounded with cold water. Take out the stopper
of the retort and pour in about 20 c.c. of strong sulphuric
acid, replace the stopper and gently heat the retort. Dark-

red fumes speedily fill the apparatus, and a heavy dark-red liquid condenses in the receiver ; when you have collected a few cubic centimetres of this liquid, remove the lamp. (Conduct the operation in a draught chamber.)   Bromine is produced in this reaction in a manner exactly analogous to that employed in preparing chlorine.

$$2KBr + 2H_2SO_4 + MnO_2 = 2Br + K_2SO_4 + MnSO_4 + 2H_2O.$$

Bromine occurs, combined with sodium, magnesium, or potassium, in sea water.   To illustrate the preparation of the pure potassium salt, from which the bromine itself can, by the above re-action, be prepared, you may perform the following experiments :—

Fig. 2L

Experiment I.—Dissolve a few crystals of potassium bromide in water in a test-tube, add a little chlorine water to the solution, and shake the tube ; the bromine is displaced from its combination with potassium by the chlorine, and is dissolved in the water; hence the yellowish-red colour now observed. Add now a few cubic centimetres of ether and shake the tube ; all the bromine is concentrated in the ethereal solution which floats on the surface ; draw this off by means of a little pipette, and add to it caustic potash until the red colour nearly disappears ; evaporate the solution on a small water-bath, formed by placing a basin on the top of a beaker containing a little water—a piece of filtering paper between the basin and the beaker separates these so that the steam may escape—(fig. 21), and when dry, ignite the residue over a Bunsen flame. You have now pure potassium bromide, from which, as you know, bromine is easily prepared.

You have already detected the very irritating odour of bromine (whence its name, βρῶμος = a bad smell).

Experiment II.—Add to a few drops of bromine about thirty times their volume of water and shake the mixture ; you thus obtain a clear yellowish-red solution. Dip a piece of madder-dyed cloth into this solution ; it is slowly bleached, the action being less intense than in the case of chlorine.

Experiment III.—Pour a few drops of bromine into a small flask, and heat it gently until the flask is full of bromine vapours ; cut and dry a small piece of phosphorus, place this on the deflagrating spoon and plunge it into the flask ; the phosphorus combines with the bromine so energetically that light is evolved. This, you remember, was also the case in a similar experiment with chlorine.

This action on phosphorus is typical of that which occurs between bromine and many other substances.

From these experiments you learn—

(1) By what method bromine is prepared.

(2) That bromine is slightly soluble in water.

(3) That this solution possesses a certain bleaching action.

(4) That the vapour of bromine is intensely irritating.

(5) That bromine has a very energetic action upon many substances.

LESSON XV.

PREPARATION AND PROPERTIES OF IODINE AND HYDRIODIC ACID.

PLACE a mixture of two grams of manganese dioxide with one gram of potassium iodide in a round-bottomed flask, resting on a retort-stand; pour about three or four cubic centimetres of strong sulphuric acid, through a funnel tube, on to the mixture, and gently heat the flask; you soon notice violet-coloured fumes appearing which gradually fill the flask, condensing in the neck, and forming brilliant metallic-looking plates. When a sufficient quantity of iodine has been thus produced, collect a little of it, by scraping it off the flask with a glass rod or horn spatula and place it in a clean dry test-tube.

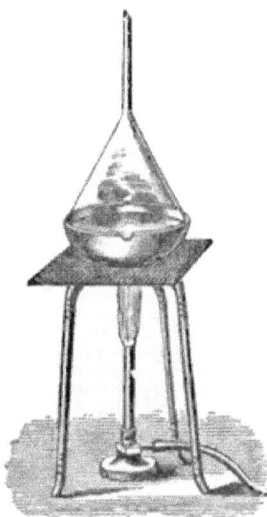

FIG. 22.

This reaction is exactly analogous to that used in last Lesson for the preparation of bromine; you can easily formulate it for yourself.

Experiment I.—Put a few crystals of iodine in a small porcelain basin (about 4 cm. diameter), place an inverted funnel over this, resting on the basin (fig. 22), and set the whole apparatus on a piece of wire gauze on a tripod stand; now very gently heat the basin, the iodine rises in beautiful violet vapours filling the funnel, on the sides of which it again condenses, forming small glancing crystals.

If you call to mind Lesson V., in which you prepared nitric acid by a process of *distillation*, you will see that this process agrees with that then performed—only you now distil a solid, by causing it to become a gas, which is again condensed to a solid : such a process is termed *sublimation.*

Experiment II.—Take a single crystal of iodine and shake it up with water in a test-tube ; very little is dissolved, the water however is slightly tinged yellow. To a few drops of this solution add a drop or two of a solution of starch in water ; a deep blue colour is instantly produced. Heat this solution ; the blue colour disappears, but reappears again when the solution cools. In the formation of this blue compound with starch you have then a means of easily detecting the presence of iodine.

Experiment III.—Make a solution of potassium iodide in a test-tube, and add to this a few drops of starch paste ; no blue colouration ensues. Now add one drop of chlorine water, the blue colour instantly makes its appearance ; on adding some more chlorine water the solution again becomes colourless, owing to the formation of a chloride of iodine which has no action on starch. Idione compounds therefore alone do not give a blue colour with starch, it is only free iodine that does this. By the addition of the chlorine water the potassium iodide was decomposed, potassium chloride being formed and iodine set free, which immediately reacted on the starch. By paying attention to what you saw in performing this last experiment, you will be able to say what special precautions must be taken in applying this test to iodine compounds.

Experiment IV.—Place a *small* piece of dry phosphorus in a basin and throw into it a few crystals of iodine ; after a very few moments combination of the two substances takes place accompanied by the evolution of light and heat.

Experiment V.—Put a few crystals of iodine into a small flask and heat this until it is filled with iodine vapours ; now throw into the flask a little piece of sodium, and continue heating the flask gently ; the sodium combines with the iodine so energetically that light is evolved.

Experiment VI.—Dissolve one crystal of potassium iodide in water, add a very few drops of a solution of potassium bromide, and then cautiously a little chlorine water; the liquid becomes yellowish-red in colour. Now add a little ether and shake the tube; the iodine, liberated from the potassium iodide by the action of chlorine, is dissolved in the ether, to which it imparts a violet colour; this ethereal solution may be drawn off by means of a pipette. A little more chlorine water is now to be added, and again a few cubic centimetres of ether; this dissolves out the bromine, which is now liberated, and which shows its presence by the distinct orange (not violet) colour imparted by it to the ether. You learn thus how iodine and bromine may be separately obtained from a liquid in which they exist together, such as the liquid produced by lixiviating kelp or the ashes of sea-plants, which again obtain their iodine and bromine from the sea itself.

Fig 23.

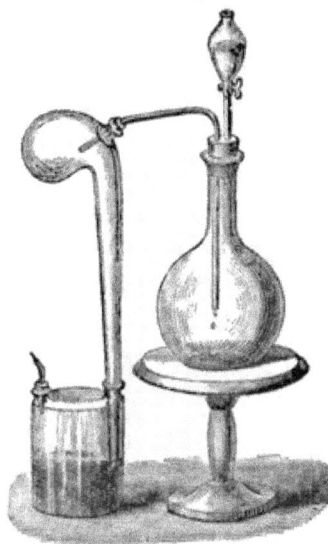

For the preparation of hydriodic acid we make use of the reaction of Experiment IV. somewhat modified.

Fit up an apparatus like that represented in fig. 23, consisting of a flask fitted with a cork carrying a small separating funnel, and a tube leading into a retort the beak of which, passing through a cork into a two-necked bottle, dips about 3 cm. below the surface of the water in this bottle (this water is coloured blue with litmus). In the flask place 2 grams of amorphous phosphorus, and 15 grams of iodine; now very gently heat the flask until these combine to form a blackish-looking mass; when this is cool, fill the bulb of the separating funnel with cold water, and, by turning the stopcock, allow this to flow, drop by drop, on to the phosphorus teriodide in the flask (this is the compound produced by the mutual action of phosphorus and iodine upon one another), which is decomposed : $PI_3 + 3H_2O = H_3PO_3 + 3HI$. Hydriodic acid, HI, is evolved, and passes into the retort, and thence into the water in the bottle, by which it is readily dissolved. This gas, being very soluble in water, is sometimes so quickly absorbed as to cause a partial vacuum in the retort, into which the water rushes ; but, before the bulb of the retort is filled with water, the level in the bottle sinks below the beak of the retort ; thus no water is allowed to flow back into the flask. The blue colour of the water you notice gradually turns to red.

Experiment VII.—Disconnect the apparatus and arrange it so that the gas shall pass into an empty dry two-necked bottle, into which, through the second neck, you lead chlorine. Violet vapours filling the bottle show the presence of free iodine ; the hydriodic acid therefore has been decomposed by the chlorine, hydrochloric acid being formed, and iodine set free. In this respect hydriodic acid differs from its chlorine-analogue hydrochloric acid.

The bromine-analogue hydrobromic acid, HBr, also exists and is formed by a reaction similar to that employed above.

From your experiments in this lesson you learn—

(1) The method of preparing iodine.

(2) The meaning of the term *sublimation* ; and that iodine may be sublimed.

(3) The reaction of free iodine with starch.

(4) The intense action of phosphorus upon iodine.

(5) The application of the reaction noted in (4) to the preparation of hydriodic acid.

(6) The solubility in water of the gas hydriodic acid, which agrees in this respect with hydrochloric acid, but differs in—

(7) The fact that it is decomposed by chlorine, hydrochloric acid and iodine being formed.

LESSON XVI.

PREPARATION AND PROPERTIES OF HYDROFLUORIC ACID, SILICON FLUORIDE, AND HYDROFLUOSILICIC ACID.

Experiment I.—Take a small circular piece of sheet lead, about 6 or 7 cm. diameter, place this in a mortar, and, with the pestle, press the centre of the lead inwards ; you thus form a little leaden cup—a platinum crucible or basin, if you have one, will do instead of this cup.   Gently warm a plate of glass large enough to cover the mouth of the little dish, and rub over it a piece of beeswax, so that the plate is equally covered with a thin coating of this substance.   Into the leaden dish put about 5 or 6 grams of powdered fluor-spar, and drench this with strong sulphuric acid.   Trace some device on the glass plate with a sharp-pointed piece of wood, or wire, making sure that the lines are drawn quite through the wax.   Place the plate over the leaden dish with the side on which the lines are drawn downwards.   Apply a gentle heat to the dish for a few moments ; you notice that white fumes are evolved and fill the space between the glass plate and the dish.   After a few minutes remove the glass plate, wash out the leaden cup, and remove the wax from the glass by rubbing it with a warm cloth ; the design which you traced is seen to be etched into the glass.   You infer there-fore that, by the action of sulphuric acid upon fluor-spar, a gas has been produced which has the property of etching glass.

You at once see the reason for not using a glass vessel in the preparation of the gas.

The reaction which goes on between the sulphuric acid and fluor-spar is thus formulated :

$$H_2 SO_4 + CaF_2 = 2HF + CaSO_4.$$

You noticed the fumes produced by the hydrofluoric acid in the air, and you have learned something about its remark-able action on glass ; the following experiments will explain this action more fully.

Experiment II.—Arrange an apparatus similar to that used in the preparation of carbon dioxide, but let the flask rest on a retort-stand ; into the flask put a mixture of 10 grams of powdered fluor-spar with about 15 grams of fine white sand. Into one of the gas bottles intended for the collection of the gas pour mercury sufficient to form a layer on the bottom to the depth of 10 or 12 mm., and let the delivery tube from the flask dip beneath this mercury (fig. 24). In the first instance, however, use a dry empty bottle for the collection of the gas.

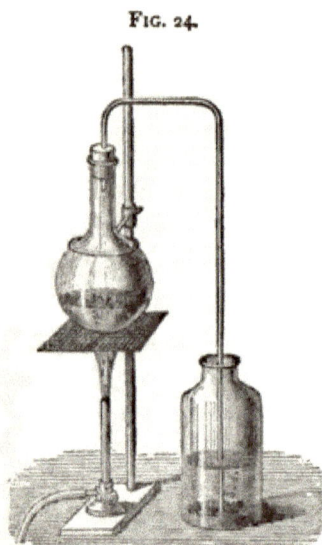

FIG. 24.

Pour, through the funnel tube, about 20 c.c. of strong sulphuric acid. Shake the flask well, so as to mix the substances in it, and apply a gentle heat. A gas, which fumes in the air, soon begins to come off. Do not apply much heat in this process, otherwise the mixture in the flask will froth up and may come over. When a few minutes have elapsed, bring a lighted taper into the gas bottle ; if it is extinguished immediately on entering the bottle, sufficient gas has been collected. You learn by this that the gas is heavier than air, that it is incombustible, and that it does not support combustion. Remove the bottle of gas, and in its place set the second bottle containing the mercury ; let this bottle be nearly filled with water, and allow the evolution of gas to proceed, while with that already collected you perform—

Experiment III.—Attach, by means of caoutchouc tubing, a small piece of glass tubing drawn to a tolerably fine point, to the end of a funnel tube. Pour a little water into the funnel ; the opening of the small tube should be of such a size that the water slowly trickles out at it. Bring this tube within the bottle containing the gas ; as each drop of water falls from the opening of the narrow tube, it seems to turn solid, and after some time you have a long stalactite hanging from the end of the tube.

Let us consider how this is brought about. Fluor-spar and sulphuric acid, as you have just learned, produce, by their mutual action, hydrofluoric acid ; this, at the moment of its production, reacts on the sand you placed in the flask (sand is nearly pure silica), and, combining with the silica, forms the gas silicon tetrafluoride, which you have collected. This gas is decomposed by water, silica being produced. As each drop of water came into contact with the silicon tetrafluoride, the gas was decomposed, and the silica thus formed coated the drop of water, holding it as in a bag :

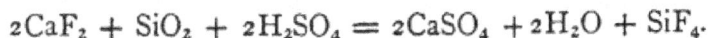

$$2CaF_2 + SiO_2 + 2H_2SO_4 = 2CaSO_4 + 2H_2O + SiF_4.$$

If you now look at the other bottle, you will see an action going forward similar to that which you have just noticed. As each bubble of gas comes up through the water, it is coated with silica, which gradually accumulates on the surface, until the water is full of this silica in a very finely divided state. Were it not for the layer of mercury the silica would soon choke up the mouth of the delivery tube. When the action has proceeded for some time, you may remove and wash out the generating flask. Cover the bottle with a glass plate, while you perform—

Experiment IV.—You remember the action of hydrofluoric acid upon glass ; you also know how this gas is produced ; and in the last experiment you have seen that the mutual action of silica, fluor-spar, and sulphuric acid, when heated together, is to produce the gas silicon fluoride : therefore you may infer that the way in which hydrofluoric acid etches glass (which is a compound of silicon), is by seizing upon the silica in the glass, and forming therewith this silicon

fluoride.   To verify this supposition is the purpose of the
present experiment.   Put a little fluor-spar into the leaden
dish, adding to it sulphuric acid as before ; cover the dish
with a glass plate, in the centre of which you have put a
drop of water, and heat gently.   You soon see that the
drop of water is coated with a film of silica ; therefore you
conclude that the gas silicon fluoride is actually produced
and is decomposed by the water.

The water in the gas bottle (Experiment II.) is now to
be separated by *filtration* from the finely divided silica
suspended in it.   Take a medium-sized glass funnel, and a
circular piece of filtering paper (fig. 25, No. 1).   (It is convenient to have your filtering paper cut in circular pieces of

FIG. 25.

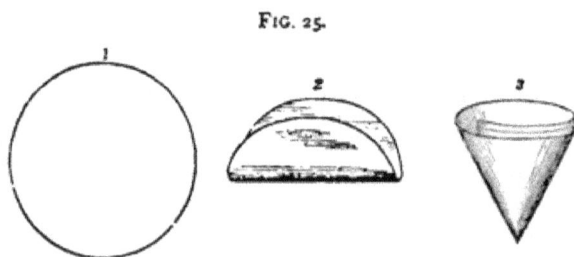

various sizes, about 4, 6, 8, 10, and 12 cm. diameter.)   Fold
the paper so as to form a half-circle (fig. 25, No. 2), and
then again at right angles to the first fold (fig. 25, No. 3).
Open out the paper thus folded, and you have a filter formed
—three folds of the paper being on one side, and one on
the other.   The funnel chosen should be rather larger than
the filter.   Set the filter inside the funnel, moisten the paper
with a few drops of water, and with the finger press the
filter to the sides of the funnel, leaving no bubbles of air
between the paper and the glass.   Place the funnel in the
ring of a retort-stand, or on a filter-stand, a beaker glass
being set below it to receive the filtered solution, or *filtrate*,
as it is termed.

Now slowly pour the liquid down a glass rod held in one hand from the gas bottle on to the filter. The liquid is thus delivered directly on to the filter, and all loss from spirting avoided. The lower end of the funnel should, for the same reason, rest against the side of the beaker ; the filtered liquid thus gently runs down the side of the glass (fig. 26). Carefully filter off the suspended silica from the liquid, set aside

FIG. 26.

the silica on the filter for further use, and proceed to examine the solution.

Touch a piece of blue litmus paper with a drop of the solution, on the end of a glass rod ; the litmus paper is reddened : you have, therefore, an acid in the solution. To a little of this acid liquid, in a test-tube, add a drop or two of barium chloride solution : an immediate white precipitate is produced. This application of *hydrofluosilicic acid* (as it is called) for the detection of barium is one which you will have occasion to use hereafter. The action of water on silicon fluoride is to form silicic acid, and hydro-fluosilicic acid :

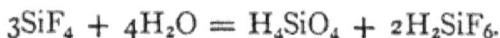

$$3SiF_4 + 4H_2O = H_4SiO_4 + 2H_2SiF_6.$$

The new acid should be kept in a gutta-percha bottle, as it gradually acts upon glass. The silica on the filter,

FIG. 27.

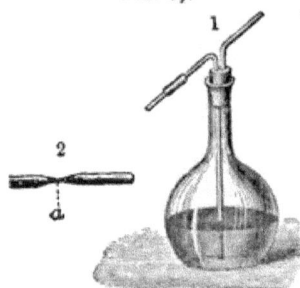

which you set aside, is now to be washed, dried, and preserved for subsequent use. To wash a precipitate, you require a wash-bottle (see fig. 27, No. 1).

Choose a flat-bottomed flask, capable of holding about three-quarters of a litre, also a good cork to fit this flask (preferably a caoutchouc cork); in the cork bore two holes. Fuse the edges of a piece of glass tubing about 10 cm. longer than the height of the flask, and bend this tube, at about 10 cm. from the end, so as to form an obtuse angle. Take another little piece of tubing, gently soften it in the Bunsen flame, keep turning it round, and *very gradually* drawing it out, so that you obtain a tube like that represented in fig. 27, No. 2, the glass in the narrow part being tolerably thick. Allow it to cool, and cut it at the point *a*. Fuse the edges of this tube, and attach it to the shorter end of that already made by a little piece of caoutchouc tubing. Take another piece of glass tubing, about 10 cm. long, bend it gently about the middle, and round its rough edges. Now fit these two tubes through the cork of the flask, so that the longer may dip to the bottom of, the shorter only a little way into, the flask. By blowing in at the shorter tube a stream of water is forced out at the narrow opening of the longer. By turning the small jet attached by flexible tubing, you can direct the stream of water in any direction you please. Fill the wash-bottle with water, and proceed to wash the silica in the filter, sending the stream of water in a circular direction round the filter, so as to bring all solid matter into the centre of the filter ; when you have filled up the filter with wash water, allow it to drain through before adding more. Continue the washings until all acid is removed (that is, until the washings no longer redden litmus paper), then cover the funnel with a piece of paper, and set it aside to dry. When dry, put the silica into a stoppered bottle, and preserve it.

In this lesson you learn—

(1) By what method hydrofluoric acid is prepared.

(2) That this acid etches glass, and the way in which it does this.

(3) By what method silicon fluoride is prepared.

(4) That this gas is decomposed by water ; and

(5) That hydrofluosilicic acid is thus produced.

(6) That a solid may be separated from a liquid by filtration.

(7) What is the use of the wash-bottle.

LESSON XVII.

PREPARATION AND PROPERTIES OF METHANE (MARSH GAS).

INTRODUCE into an iron tube about 20 cm. long and 5 cm. diameter (see fig. 28) a mixture of 8 grams of dry sodium acetate with 8 grams of caustic soda—previously strongly heated on an iron plate—and 12 grams of lime. Fit in a good cork with a delivery tube, arrange the trough and a bottle to collect the gas ; now heat the tube, beginning at the

F

upper part, and heating gradually downwards. As the gas does not come off until a pretty high temperature is reached,

Fig. 28. it is preferable to use such an apparatus as that described, than to heat the mixture in a glass flask, which is very liable to crack.*

$$NaC_2H_3O_2 + NaHO = Na_2CO_3 + CH_4.$$

The lime in the above mixture serves to render it more infusible, and to prevent it stopping up the tube. As the bubbles of gas pass through the water in the trough, test them from time to time by bringing a lighted taper near them. As soon as they inflame, begin to collect the gas; and when you have filled three bottles, withdraw the delivery tube from the water.

Experiment I.—Set an inverted bottle of the gas on the ring of a retort-stand, and apply a light to its mouth; the gas burns with a bluish non-luminous flame.

Experiment II.—Pour a little lime-water into the bottle used in the last experiment, when the flame has gone out; on shaking the bottle you learn, from the turbidity in the lime-water, that carbon dioxide is present. One of the products of the oxidation or burning of marsh gas is therefore carbon dioxide.

Marsh gas is formed in coal mines, where it is known under the name of 'fire-damp.' If mixed with ten volumes of atmospheric air it forms an explosive mixture, producing water and carbon dioxide. This latter gas will not support life; hence the dreadful effects produced, not only by the actual explosion in mines, but also by the formation of this 'choke' or 'after' damp, as the carbon dioxide is called by the miners.

Experiment III.—To illustrate the explosive nature of a mixture of marsh gas and oxygen, decant into a soda-water

* This gas may also be easily prepared by heating 1 part of dry sodium acetate with 2 parts of sodium carbonate and 2 parts of lime.

bottle one volume of the gas, and add two volumes of oxygen; cork the bottle, shake it, and wrap a cloth tightly round it, then take out the cork, and bring a lighted taper to its mouth; a very sharp explosion ensues.

Experiment IV.—Fill a perfectly dry flask, capable of holding about 250 c.c., with marsh gas, by upward displacement, and fill also a similar flask with chlorine; connect these flasks by a glass tube passing through corks in each, and set them in the sunlight, the flask containing marsh gas being uppermost (see fig. 29). You soon notice that the upper as well as the lower flask becomes filled with yellow fumes, but that these after a little time entirely disappear. Now separate the flasks, and bring a bottle containing strong ammonia solution near the mouth of each; dense white fumes are formed: this, you remember, (Lesson IX.), tells you that hydrochloric acid is present. The chlorine has therefore withdrawn part of the hydrogen from the marsh gas. Another quantity of chlorine combines with the remainder of the marsh gas, forming a new gaseous compound, the presence of which you may detect by its peculiar ethereal odour after agitating the gaseous mixture in the flask with a little water to absorb the hydrochloric acid.

FIG. 29.

You learn then from these experiments—

(1) By what method marsh gas may be prepared.

(2) That this gas is combustible.

(3) And that by its combustion carbon dioxide is produced.

(4) What is the cause of the evil effects produced by this marsh gas when present in mines.

(5) That chlorine acts on marsh gas in sunlight.

LESSON XVIII.

PREPARATION AND PROPERTIES OF ETHENE (OLEFIANT GAS).

FIT up an apparatus like that shown in fig. 30, consisting of two flasks, the first of which is fitted with a cork carrying a tube bent twice at right angles ; the last limb of this tube dips, through a wider tube, into the second flask, from which a delivery tube leads to the trough. The second flask

FIG. 30.

also carries a thermometer, as shown in the figure.   Into the first flask pour about 20 c.c. of alcohol (rectified spirits). Mix 40 c.c of strong sulphuric àcid with 12 c.c. of water, and pour this mixture into the second flask, and fit the corks into their places.   Heat both flasks; the alcoholic vapours pass into the sulphuric acid (the temperature of which must not be raised above 170°), the acid gradually darkens in colour, and bubbles of gas come up through the water in the trough.   When, on bringing a light close to it, the issuing gas takes fire, you may begin to collect it.   Fill three bottles with ethene gas, withdraw the lamps, and at once take the delivery tube of the first flask out of the liquid in the second by drawing it up the wider tube. In this reaction the sulphuric acid acts in a manner similar to that noticed in the case of oxalic acid, viz. it withdraws the elements of water:

$$C_2H_6O = C_2H_4 + H_2O.^*$$

Experiment I.—Apply a light to the mouth of the first bottle standing on the table; the gas burns with a bright white smoky flame, while the taper is extinguished. Ethene is therefore a combustible gas. After the combustion is over, add a little lime-water to the bottle, and shake it briskly; the turbidity of the liquid indicates the presence of carbon dioxide.

Experiment II.—Graduate a soda-water bottle into four divisions (see Lesson III.), and decant into it one measure of ethene gas, and three measures of oxygen; cork the bottle, surround it with a cloth, and shake it several times; on withdrawing the cork and applying a light, a loud explosion occurs You have in this experiment completely oxidised the carbon and hydrogen of the ethene, carbon dioxide and water being the products:

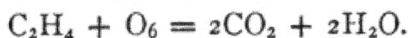

$$C_2H_4 + O_6 = 2CO_2 + 2H_2O.$$

Experiment III.—Pour a few drops of bromine into a small flask, shake the flask, and pour the excess of bromine back into the bottle; and bring the flask, mouth downwards, into one of the bottles of ethene. The red vapours quickly disappear, while heavy oily drops form on the sides of the bottle, and run down to the bottom. This oil is a compound of ethene with bromine. A similar oily compound is produced by the union of this gas with chlorine; hence the name *olefiant* or *oil-producing* gas sometimes given to ethene.

---

* This reaction probably occurs in two stages; in the first ethyl sulphuric acid is formed, and this is then split up into ethene and sulphuric acid.

Experiment IV.—Graduate a gas bottle into three equal parts, fill it with water at the trough, and quickly decant up into it two measures of chlorine, and one of ethene ; cover the mouth of the bottle with a glass plate, remove it from the trough, agitate it briskly, and apply a light ; the gases burn with a very smoky flame, which gradually passes down the bottle, much carbon being deposited. This experiment further illustrates the great affinity which chlorine has for hydrogen ; the ethene is robbed of its hydrogen and the carbon is deposited on the glass.

In this lesson you have learned—

(1) What is the action of strong sulphuric acid upon alcohol.

(2) By what mode ethene is prepared.

(3) That ethene is a combustible gas, burning with a smoky flame. (Coal gas owes its luminosity in great measure to the presence in it of ethene.)

(4) That for its complete oxidation or burning, one volume of ethene requires three volumes of oxygen.

(5) That with chlorine and bromine ethene forms oily compounds.

(6) But that with different proportions of this gas and chlorine the hydrogen of the former is seized upon by the latter, carbon being liberated. Contrast this with the action of chlorine upon methane (last Lesson), whereby a *substitution* product of the methane (that is, a product in which part of the hydrogen is replaced by another element) and hydrochloric acid are produced.

(7) That the explosion which occurs when a light is brought into a mixture of coal gas and air is caused by the combination of the methane and ethene, contained in the gas, with the oxygen of the air.

LESSON XIX.

PREPARATION AND PROPERTIES OF ETHINE (ACETYLENE).

THIS gas is produced in many cases of incomplete combustion of organic substances ; thus, when a Bunsen lamp burns down, the gas is only partially consumed, ethine is produced, and can be recognised by its peculiar smell. This gas combines with certain metals, as copper, &c. The compound produced when ethine is passed through an ammoniacal solution of cuprous chloride is very characteristic, possessing a bright red colour ; we make use of this reaction in order to detect the presence of ethine.

Experiment I.—Make a solution of cuprous chloride by dissolving 10 grams of black cupric oxide in about 100 c.c. of ordinary hydrochloric acid, boil for 15 minutes with 8 grams of metallic copper in small pieces, pour this solution into about one litre of water, allow the precipitate to subside, pour off the water, and rinse the precipitate into a bottle,—of about 150 c.c. capacity,—nearly full of water. After the subsidence of the precipitate, the water is again poured off, 40 grams of powdered ammonium chloride are added, the bottle is again filled with water and shaken up.

To a small quantity of this solution gradually add ammonia until the precipitate which forms redissolves, and you have an azure-blue solution. Pour a few cubic centimetres of this solution into a wide-necked flask, capable of holding about half a litre, rinse the flask with the liquid and allow it to drain for a few moments ; turn down a Bunsen burner so that it may *burn below*, and now bring the flask, mouth downwards, a little way over the top of the lamp, and support it in this position on a retort-stand ; after a few minutes you will notice that the inside of the flask becomes coated with a red substance, which is the compound produced by the action of ethine upon cuprous chloride solution. It is termed cuproso-vinyl oxide, and has the formula $(C_2HCu_2)_2O$.

To show the presence of ethine in coal gas, we will perform—

Experiment II.—Choose a U-tube each limb of which is about 12 or 15 cm. long and 1 cm. diameter, and fit into each of the openings a cork carrying a little piece of glass tubing ; let one of these tubes be drawn to a tolerably fine opening, to the other attach a caoutchouc tube connected with the gas tap ; pour into the U-tube sufficient ammoniacal cuprous chloride solution to fill the bend, support the tube on a clamp or retortstand and turn on the gas, which, bubbling through the solution in the bend of the tube, issues from the small glass tube, where it may be lighted (fig. 31) ; after a few minutes you notice the formation of a copious red precipitate in the blue solution, which proves the presence of ethine in the coal gas.

FIG. 31.

You have learned—

(1) That ethine is produced in very many cases of incomplete combustion.

(2) That ethine is present in coal gas.

(3) How to detect ethine.

LESSON XX.

STRUCTURE OF FLAME.

FROM the experiments performed in Lesson III. you have learned the exact relation which exists between the terms 'combustible' and 'supporter of combustion,' and from various experiments you have learned what is meant by the term 'combustion.'

In Lesson XVIII., Experiment I., you saw that ethene burns with a smoky, *luminous* flame.

Experiment I.—Set up the apparatus used in preparing ethene (fig. 30, p. 68), putting about 10 c.c. of alcohol and a corresponding amount of sulphuric acid into the flasks; to the end of the delivery tube adapt, by means of caoutchouc tubing, a small glass tube having a tolerably fine orifice. Fill a large wide-mouthed bottle with oxygen; when ethene gas is coming off in a gentle stream, light it; it burns with a smoky flame. Now bring the bottle of oxygen gas, mouth downwards, over the burning gas; the flame becomes much less luminous and more elongated. You thus see that when this gas is completely burned, or oxidised, the luminosity of the flame is diminished.

Experiment II.—Fit a two-necked Wolff's bottle with corks, through one of which passes a small glass tube drawn to a tolerably fine point, and through the other a chloride of calcium tube, in the bulb of which is a little cotton-wool which has been previously soaked in benzene. In the bottle place materials for generating hydrogen (see fig. 32). When the gas has been coming off for a little time, test it by filling a test-tube by upward displacement and bringing this near a flame; if the hydrogen burns quietly in the tube, you know that the air has been driven entirely out of the bottle. Now light the gas issuing from both necks of the bottle. The flame of the hydrogen burning at the mouth of the small glass tube is almost colourless, while that at the

mouth of the other tube is tolerably luminous. The gas
giving the latter flame passes, before being burned, through

FIG. 32.

wool soaked in benzene, which is a
hydrocarbon, or compound of carbon
and hydrogen, resembling in this respect ethene. But the presence of
heavy hydrocarbons in a flame gives
luminosity to that flame. On the other
hand, a flame in which hydrocarbons
of low density only are present is
wanting in luminosity. The luminous
character of the flame of coal gas is
now explained, as you already know
that that gas contains hydrocarbons of
comparatively high density. It is also
not improbable that in the burning of
such luminous flames a certain amount of carbon is liberated, and that this also radiates light.

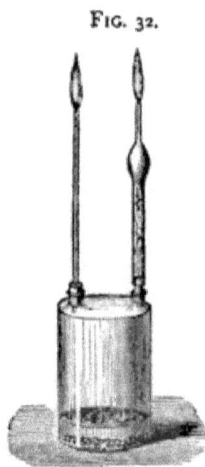

Experiment III.—Fill two bottles with chlorine by
downward displacement, and bring one of these over a
Bunsen lamp which is burning with a small flame ; the luminosity of the flame is much increased.

You have learned that chlorine combines with ethene,
methane, &c. (components of coal gas), and also that chlorine has a great affinity for hydrogen. Both of these actions
probably occur in the present case, and hence, on account
of the density of the products of combustion, as well as, pos-
sibly, on account of the presence of solid carbon, the luminosity of the flame is much increased.

Experiment IV.—Set up a small hydrogen-generating
apparatus having a platinum jet adapted to the end of the
delivery tube at which the issuing gas is ignited. On bringing the second bottle of chlorine over the burning hydrogen,
an increase in the luminosity of the flame is noticed. The
product of combustion of hydrogen in chlorine—hydrochloric acid—possesses a density twice as great as that of the

product of combustion of hydrogen in air, viz. water ; therefore the flame appears brighter when the bottle of chlorine is
brought over the burning hydrogen.   In this case the presence of solid matter in the flame is impossible.

Experiment V.—Bend a glass tube, about 60 cm. long,
into the shape shown in fig. 33, No. 1; bring the end *a* into
the flame of a candle immediately above the wick ; on applying a light to *b*, you find that combustible gases are being
withdrawn from the flame.

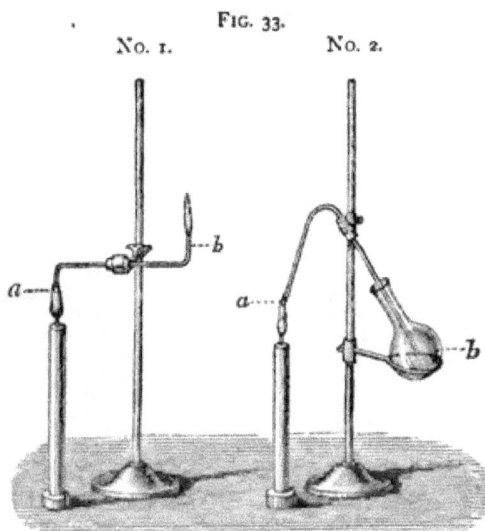

FIG. 33.

No. 1.   No. 2.

Experiment VI.—Move the end *a* of the tube until it is
situated at the outside of the candle flame; a light now brought
to the end *b* is extinguished, telling you that a non-combustible
gas is present   Allow the end *b* of the tube (fig. 33, No. 2) to
dip into a small flask, into which, after a little time, pour a
few cubic centimetres of lime-water ; on shaking the bottle
a precipitate forms in the solution.   The gas which you have
drawn off is therefore carbon dioxide. These experiments tell
you that in the flame of a combustible carbonaceous substance there is an inner zone in which the gases are but
partially burned, then a zone containing carbonaceous matter

(constituting the luminous part of the flame), while outside of this is the zone of complete combustion in which all the carbon is burned to carbon dioxide. The presence, in this part of the flame, of the product of combustion of the hydrogen, viz. water, may be shown by bringing a test-tube, containing a little cold water, into the outer flame of the candle; drops of dew are immediately formed on the tube, by the condensation of the water vapour on the cold glass.

There is yet another part of the flame, called the *mantle*, which is situated at the extreme outer edge of the flame,

FIG. 34.

FIG. 35.

FIG. 36.

and only becomes visible by holding a piece of cardboard (cut so as exactly to cover all that part of the flame ordinarily visible) between the eye and the flame (fig. 34); the mantle will then be seen round the outer edge of the cardboard.

Experiment VII.—Make a small helix of copper wire (see fig. 35), leaving a space between each coil, and bring this over the flame of a candle; the flame is extinguished. As air sufficient to support combustion could enter between the coils of wire, the flame could not be extinguished by suffocation.

Experiment VIII.—Pour, from a small basin grasped by a pair of crucible tongs, a quantity of burning methylated spirit on to a piece of wire gauze held over a second basin (fig. 36). The spirit passes through the gauze, but it is no longer ignited. The gauze presents to the burning spirit a large metallic surface, which conducts away heat from the flame, so that the process of rapid oxidation of the constituents of the spirit ceases, and the flame therefore goes out. The same thing occurred in Experiment VII. The copper wire so quickly conducted away heat from the candle flame, that the temperature required to inflame the mixture of gases could no longer be maintained. On this principle of extinguishing a flame Sir Humphry Davy based his celebrated ' safety lamp.'

In this lesson you have learned—

(1) What is the difference between a luminous and a non-luminous flame.

(2) What is the constitution of a candle flame, as the type of others.

(3) That a flame may be extinguished by heat being conducted from it by a metallic surface.

LESSON XXI.

PROPERTIES OF SULPHUR.

Experiment I.—Heat a quantity of 'flowers of sulphur' in a Florence flask placed on wire gauze on a retort-stand. You notice that the sulphur melts, forming a clear mobile liquid, it then darkens and gets thicker, and, as the heat is increased, it again becomes fluid. When this point is reached, take hold of the flask with a cloth or pair of crucible tongs, invert it, and pour the molten sulphur into cold water in a beaker. The mass of sulphur in the water retains its plasticity, and can be moulded with the hand, like recently melted caoutchouc.

When sulphur is allowed to spontaneously crystallise

from its solution in carbon disulphide, it forms *octahedral*

FIG. 37.

crystals of forms derived from that shown in fig. 37 ; but, as you will learn from the next experiment, it may also be obtained in another crystalline form.

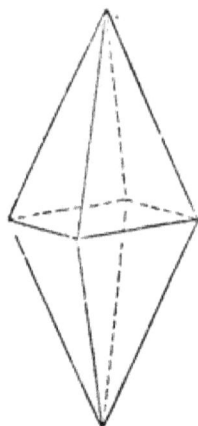

Experiment II.—Melt 20 grams of sulphur in a Hessian crucible placed in a fire, and allow it to cool until a crust forms on the surface of the mass ; now make two holes in this crust with a glass rod, and invert the crucible so that the still molten sulphur flows out at one of the holes into a dish of water, while air enters at the other. On removing a portion of the sulphur with a knife, you will find that the inside of the crucible is full of *needle-shaped* or *prismatic* crystals of sulphur.

You have already learned that sulphur burns in air or oxygen, producing the gas sulphur dioxide.

Experiment III.—Fuse a small quantity of sulphur, mixed with an equal quantity of sodium carbonate, in a porcelain crucible over a Bunsen flame, when cool pour a little water into the crucible and dip a silver coin into the solution. The silver is stained black owing to the production of silver sulphide. This reaction is occasionally used for the detection of sulphur.

From these experiments you learn—

(1) What is the action of heat upon sulphur.

(2) By what method the presence of sulphur may be detected.

(3) That sulphur is capable of existing in various modifications, viz. as plastic, octahedral, or prismatic. Such modifications of one and the same substance are called *allotropic* modifications of that substance.

LESSON XXII.

PREPARATION AND PROPERTIES OF SULPHUR DIOXIDE AND
TRIOXIDE.

INTO the flask of an apparatus similar to that represented in
fig. 38, put about 20 grams of copper clippings (take care to
slide the pieces of copper down the side of the flask to prevent cracking it), and add, through the funnel tube, 60 c.c.
of strong sulphuric acid. Set the flask on a retort-stand,

FIG. 38.

and let the delivery tube dip into a gas bottle. Place a
Bunsen lamp beneath the flask and gradually increase the
heat; after a little while you notice a brisk action going on in
the flask; when the action has thus begun, you may moderate
the heat. You soon perceive that the gas which is given off
has a very pungent sulphurous odour. It is identical with
that produced by burning sulphur in oxygen, viz. sulphur

dioxide : $2H_2SO_4 + Cu = CuSO_4 + 2H_2O + SO_2$. Collect two bottles of the gas, and while the evolution of the gas is proceeding, get ready an apparatus in which to *condense* the sulphur dioxide. Choose a piece of stout glass tubing about 2 cm. diameter ; at a distance of 20 cm. from one end soften the tube in the flame (in the case of thick tubing such as this, it is better to use the blowpipe, *very gradually* increasing the heat of the flame), and draw it out; by heating the projecting portion in the flame, touching it with a little piece of glass-rod, and sharply drawing it off, you will get rid of most of this part ; finally, heat the end of the tube to redness, keep turning it round, and blow in at the open end once or twice; you will thus get a tube sealed neatly at one end.

FIG. 39.

Now heat the tube at about 8 cm. or so from the open end, and gently draw it out until the narrow part is just a little wider than the delivery tube of the gas-generating apparatus (fig. 39). Select a beaker whose height is about equal to the length of the tube you have just prepared; set this, with its open end upwards, in the middle of the beaker, and surround it with a freezing mixture made by intimately mixing together one part of pounded ice with about one and a half parts of salt.

By this time the two gas bottles will be full of sulphur dioxide, cover them with greased glass plates, and bring the delivery tube into the wide tube standing in the freezing mixture, so that it may reach nearly to the bottom of this latter tube. Allow the evolution of gas to proceed for some time ; meanwhile examine that which you have collected.

Experiment I.—Plunge a lighted taper into one of the bottles ; the flame is extinguished, and the gas does not take fire ; quickly replace the glass plate.

Experiment II.—Plunge a few flowers (preferably red roses or reddish-coloured pansies) into the same bottle of sulphur dioxide. The red colour of the flowers is after a little time discharged. This gas, therefore, possesses the property of bleaching ; further experiments will explain this more fully.

Experiment III.—Invert the second bottle beneath water in the trough, and withdraw the glass plate; the water rises in the jar, showing that the gas is soluble in water. Slip the hand beneath the mouth of the bottle, remove it from the water, and shake it briskly; again bring it under the water and repeat the operation until the water has nearly filled the bottle. This solution contains *sulphurous acid*—H_2SO_3 = $SO_2 + H_2O$.

Experiment IV.—Transfer a few cubic centimetres of this solution to a test-tube, and add a drop or two of barium chloride solution; no change ensues. Boil a second portion of the sulphurous acid solution with a few drops of nitric acid, and again add barium chloride; the immediate formation of a white precipitate shows that there is a difference between this and the original liquid. This white precipitate with barium chloride, as you will afterwards more fully learn, tells us that *sulphuric acid* is present. The sulphurous acid has, there-fore, been changed to sulphuric acid, i.e. it has taken up oxygen, $H_2SO_3 + O = H_2SO_4$. It is in this way, viz. by robbing substances of their oxygen, or acting as a *reducing agent*, that sulphur dioxide bleaches.

Experiment V.—To a few cubic centimetres of sulphurous acid solution add a little bromine water and boil the mixture. On applying the test for sulphuric acid—viz. barium chlo-ride—to a portion of the liquid, the formation of a white precipitate tells you that this acid has been formed.

The action which has here taken place is analogous to that noticed in the last experiment. The oxygen of the water has been seized upon by the sulphur dioxide, while the hydrogen which remains has combined with the bromine to form hydrobromic acid. To prove the presence of the latter substance, add a little silver nitrate solution to another part of the liquid ; an immediate white precipitate of silver bromide is produced.

Experiment VI.—Pour a small quantity of a solution of potassium iodide into a beaker, add some water and a few drops of starch paste, then one or two drops of chlorine water ; as you have already learned, an intense blue-coloured solution is thus obtained. To this solution add, drop by drop, with constant stirring, sulphurous acid solution until the blue colour is entirely discharged. This bleaching is effected by the sulphurous acid combining with the oxygen of the water to form sulphuric acid, while the hydrogen of the water combines with the iodine liberated from the potassium iodide to form hydriodic acid, which, as you know, is colourless, soluble in water, and without action on starch.

You thus learn that sulphur dioxide bleaches in a way directly opposed to that in which chlorine bleaches ; the former acts by *reduction*, i.e. by taking away oxygen ; the latter by *oxidation*, i.e. by giving oxygen.

You must now turn your attention to what is going forward in the tube in the freezing mixture. If a steady flow of gas has been kept up, you will find, on raising the tube partially out of the beaker, that it contains a liquid ; by the application of cold you have caused the gaseous sulphur dioxide to become liquid. Many gases can be thus condensed, while others refuse, under all circumstances, to appear in any other than a gaseous form. The latter are termed, on this account, *permanent gases*. Keeping the tube as much as possible in the freezing mixture, direct a blowpipe flame on to the narrowed portion ; when fused draw this out, but do not quite close it ; allow the tube to get quite cold, again heat the narrow part, and now quickly draw it off. You may remove the tube from the beaker and set it aside ; the sulphur dioxide will remain liquid in the closed tube.

Experiment VII.—Heat a few drops of the sulphurous acid solution in a test-tube. You distinctly smell that sulphur dioxide is given off. Let us try to oxidise this sulphur dioxide, and thus produce the higher compound of sulphur with oxygen, viz., sulphur trioxide.

Experiment VIII.—Pass a stream of sulphur dioxide into a part of the solution of sulphurous acid, until the liquid smells strongly of the gas; then transfer it to a small flask, fitted with a cork, in which two holes are bored; through one of these passes a tube, which reaches nearly to the bottom of the flask, while through the other a shorter tube is fitted which ends just below the cork.

Fit up a small oxygen-generating apparatus, but let the

FIG. 40.

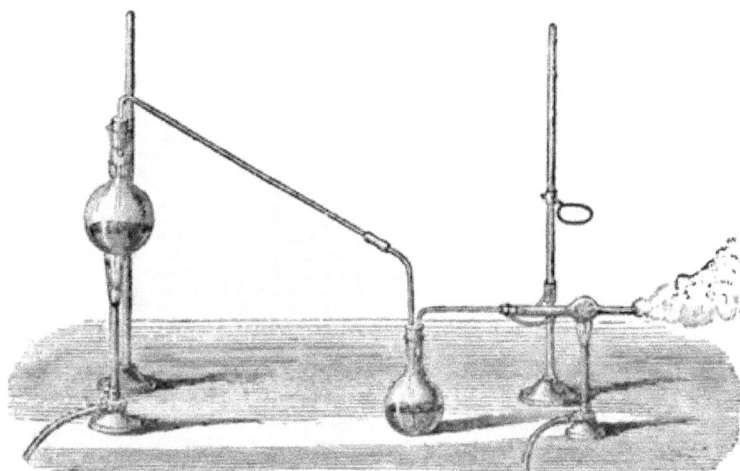

end of the delivery tube be bent at a right angle with the longer part of the tube. Support this on a stand, and con- nect it, by caoutchouc tubing, with the longer tube coming from the flask. Take a few pieces of asbestos, soak them in platinum tetrachloride solution, then dry and ignite them in a small basin over a Bunsen lamp. The platinum tetra- chloride is decomposed, chlorine being given off, while platinum remains behind in a finely divided state covering the asbestos, which therefore appears nearly black. Put this platinised asbestos into the bulb of a chloride of calcium tube, and through a cork in the wider end of this tube pass the short tube leading from the flask. The apparatus is represented in fig. 40.

Cause a gentle stream of oxygen to pass through the solution of sulphurous acid in the flask; the oxygen carries along with it some of the sulphur dioxide, the presence of which is easily distinguished by the smell of the gases issuing from the bulb tube. On gently heating the bulb containing the platinised asbestos, the two gases, oxygen and sulphur dioxide, are caused to unite. Sulphur trioxide, SO_3, is hereby produced, the presence of this compound being shown by the dense white fumes which now appear near the orifice of the bulb tube. This property of fuming in the air is characteristic of the trioxide.

By means of caoutchouc tubing, adapt to the end of the bulb tube a small glass tube, bent at right angles, and allow this to dip into a little water in a beaker. Continue to heat the bulb, and condense the resulting sulphur trioxide in the water. After a little time test some of this water with barium chloride; you find that it contains sulphuric acid. You have therefore produced this acid by the combination of sulphur trioxide with water :

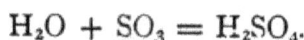

$$H_2O + SO_3 = H_2SO_4.$$

Experiment IX.—Pour about 10 c.c. of strong Nordhausen sulphuric acid ($H_2S_2O_7$) into a little retort, to which a small receiver is attached. On heating the acid in the retort, fumes of sulphur trioxide are formed, which pass over into the receiver, where they condense to a white crystalline solid :

$$H_2S_2O_7 = H_2SO_4 + SO_3.$$

If the receiver be carefully closed, these crystals of sulphur trioxide may be kept for a long time. Allow a drop of water to fall on to one of the crystals : the chemical combination is so intense, that it hisses as if it had touched red-hot iron. You already know that sulphuric acid is the product of this reaction.

From the experiments in this lesson you learn—

(1) What are the methods for preparing the two oxygen compounds of sulphur, viz., the di- and tri-oxide.

(2) What is the way in which sulphur dioxide bleaches.

(3) What is the action of this gas on a solution of iodide of starch. This action, as you will hereafter learn, constitutes the basis of a widely applicable quantitative process.

(4) That sulphur dioxide may be liquefied; and the meaning of the term *permanent gas.*

(5) That sulphur trioxide is the product of the oxidation of sulphur dioxide, and that this compound, in contact with water, yields sulphuric acid.

LESSON XXIII.

PROPERTIES OF SULPHURIC ACID.

To make sulphuric acid by the process adopted on the manufacturing scale requires a more complicated arrangement of apparatus than is readily put together in the laboratory. We shall, therefore, neglect the actual preparation of this substance, and proceed to perform a few experiments with the acid, which is a liquid in constant use in the laboratory.

Experiment I.—To a drop or two of sulphuric acid diluted with water, in a test-tube, add barium chloride solution ; a copious white non-crystalline precipitate is at once produced. After boiling, allow the precipitate to settle, pour off the clear liquid, add a little water to the precipitate, shake it up, and again allow it to settle. After having thus washed it several times *by decantation*, pour over the precipitate a little strong hydrochloric acid; it remains undissolved. Pour away the hydrochloric acid, and add a little nitric acid; the precipitate is unacted upon. Therefore the formation of a white precipitate (barium sulphate), *insoluble in acids*, when barium chloride is added to any given solution, tells us that sulphuric acid or a sulphate is present in that solution.

Experiment II.—Heat a little of the acid, and pour it on to a small piece of sugar in a porcelain basin. The sugar is instantly charred; steam is given off, while black masses of charcoal remain. Sulphuric acid chars sugar, which has the composition $C_{12}H_{22}O_{11}$ by removing the elements of water, viz., hydrogen and oxygen, and eliminating the carbon.

You thus learn—

(1) In what way sulphuric acid may be recognised.

(2) That this acid has a great avidity for water.

LESSON XXIV.

PREPARATION AND PROPERTIES OF SULPHURETTED HYDROGEN.

FIT up an apparatus similar to that shown in fig. 41, and in the flask A place a few small pieces of ferrous sulphide. Cover these with water, and pour a little strong sulphuric acid through the funnel tube into the flask. Sulphuretted hydrogen gas is evolved, and is washed by passing through water contained in the small flask B. The reaction is as follows: $FeS + H_2SO_4 = FeSO_4 + H_2S$. Allow the exit

tube to dip into a dry gas bottle; place the whole apparatus in the draught chamber, and continue the evolution of
the gas for some time. Now close the bottle with a well-
greased glass plate, remove it, and place in its stead a
beaker containing water.

Experiment I. — With-
draw the glass plate, and
bring a lighted taper to the
mouth of the bottle con-
taining the gas. The gas
burns with a pale bluish
flame, and the inside of the
bottle becomes coated with
sulphur. In this reaction
sulphur dioxide and water
are produced together with
the sulphur.

Fig. 41.

$$H_2S + O_3 = H_2O + SO_2 \quad \text{and}$$
$$H_2S + O = H_2O + S.$$

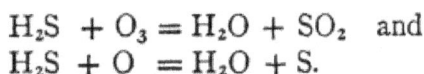

If the gas be mixed with air in quantity sufficient to burn
it completely, the first reaction alone takes place.

The water in the beaker glass will now be saturated with
the gas, the peculiar smell of which it will possess.

Experiment II.—Prepare solutions of copper sulphate,
arsenious oxide, ferric chloride, and barium chloride in
water, acidulate them slightly with hydrochloric acid, and
add to each a little of the solution of sulphuretted hydrogen.
In the first liquid, a black precipitate of copper sulphide
(CuS) will be produced ; in the second, a yellow precipitate
of arsenious sulphide (As_2S_3) ; in the third, a finely divided
precipitate of sulphur ; the fourth liquid will remain clear.
By taking advantage of the reaction of metallic solutions towards sulphuretted hydrogen, we are enabled to arrange the
metals in groups for the purposes of qualitative analysis.
(See p. 106.)

LESSON XXV.

PREPARATION AND PROPERTIES OF PHOSPHORETTED HYDROGEN.

FIT a cork carrying an exit tube into a small round-bot-tomed flask, supported on a clamp, in which you have already placed a strong solution of caustic potash and a few pieces of phosphorus. Connect the end of the exit

FIG. 42.

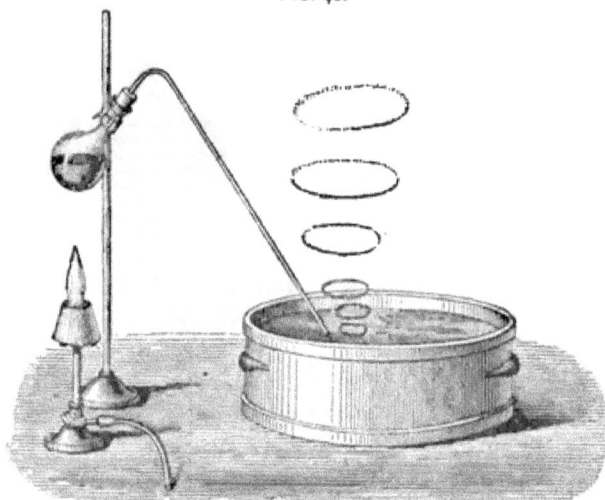

tube, by means of a caoutchouc tube, with the gas tap, and pass coal gas into the apparatus, the cork being loosely held in its place, until it is entirely filled with it. Now shut the gas stopcock, remove the caoutchouc tube, fit the cork tightly into the flask, and quickly plunge the end of the exit tube under water in a pneumatic trough. On now heating the flask a gas is evolved, each bubble of which, as it passes out of the water into the air, spontaneously inflames, and produces a ring of white smoke (see fig. 42). The reaction which takes place may be thus formulated :

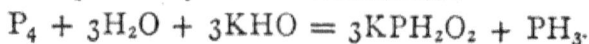

$$P_4 + 3H_2O + 3KHO = 3KPH_2O_2 + PH_3.$$

The rings are composed of phosphoric anhydride, formed by the oxidation of the phosphoretted hydrogen ($2PH_3 + O_8 = P_2O_5 + 3H_2O$).

In this lesson you learn by what method phosphoretted hydrogen is prepared, and also that it is spontaneously inflammable.

PART II.

QUALITATIVE ANALYSIS.

QUALITATIVE ANALYSIS is that branch of Practical Chemistry which treats of the methods of determining the nature of substances, and the way in which their constituents may be separated. By far the greater number of substances may be resolved, by various forces, into parts or constituents, which separately are possessed of properties different from those which characterise the original matter. These parts may, generally, be again broken up ; but ultimately we arrive at certain forms of matter which refuse to yield to any force by means of which we may attempt to subdivide them.

These ultimate parts we term *Elements* ; by their union in definite proportions they form *Compounds*. For example, a piece of chalk (calcium carbonate) when heated is split up into two groups or combinations of elements—carbon dioxide, which is evolved ; and lime, which remains behind. By appropriate means we can further resolve the carbon dioxide into carbon and oxygen, and the lime into calcium and oxygen ; but all attempts to break up the calcium, the carbon, or the oxygen into simpler substances have hitherto been unsuccessful : these three substances, calcium, carbon, and oxygen, are therefore called elements. If, then, we desire to know of what substances a given piece of matter is composed, our wish will be satisfied when we have ascertained what elements, or what combinations of elements, are present in it.

The total number of the elements at present known to us is about 63, but comparatively few of these are of common occurrence, and still fewer have received any practical application.

The operations of Qualitative Analysis may be conveniently subdivided under the two heads of (1) Analysis by dry reactions; and (2) Analysis by wet reactions.

In determining the nature of a substance by the dry method, we subject it to examination when in the solid state ; for example, we notice if it imparts any peculiar colour to flame ; if it yields a metal when heated with reducing substances at a high temperature; or if, when heated, it imparts a colour to various fluxes, &c.

In analysing a substance in the wet way, we subject it to the action of appropriate solvents, and determine the nature of the substances dissolved by the addition of reagents, themselves also in solution.

SECTION I.

GENERAL PRELIMINARY OPERATIONS.

A. *FLAME REACTIONS.*

THE flame of the Bunsen lamp may be used to detect the presence of very many elements.

The Bunsen lamp consists essentially of a metal tube, at the bottom of which a gas burner is fixed ; the lower part of this tube is pierced with holes, through which air enters and mixes with the gas in the tube. These holes are of such a size as to admit an amount of air, which, when mixed with the gas issuing from the burner, is sufficient to oxidise or burn it entirely. A non-luminous, very hot flame is thus produced.

The Bunsen lamp to be used in the following experiments is shown in fig. 43, No. 1. At *a* is a circular cap, by moving which you can regulate the supply of air, so as to obtain a more or less luminous flame. A conical chimney, resting on the gallery (*b*) of the lamp, serves to protect the flame from draughts of air.

The flame is shown in fig. 43. No. 2, of its natural size ; the letters refer to the various portions of the flame, with the properties of which it is necessary that you make yourself thoroughly acquainted.

There are three principal divisions, viz.—

(1) The dark zone, *a a, a a*, in which the cold, unburnt gas is mixed with about 62 per cent. of air.

(2) The flame mantle, *a, ca, b*, in which is a mixture of burning coal gas and air.

(3) The luminous point *aba*, which is not seen when the full supply of air is allowed to enter the lamp, but which can be produced by turning the cap, so as partially to close the holes at *a*, No. 1.

Examining the flame more in detail, we find the following six points, which are made use of in the reactions :—

1. The base of the flame is situated at *a*, a very small distance from the summit of the lamp itself. On account of the proximity of the metal tube, and also by reason of the ascending current of cold air, the temperature of this part of the flame is, comparatively, very low. Hence many easily volatilised substances may be recognised by the colours which they impart to the flame when held in this portion of it.

2. The zone of fusion, or point of highest temperature, is situated at β, somewhat above the first third of the entire height of the flame, and at a distance from the edge of the flame equal to about one-fourth of its greatest breadth.

3. and 4. The lower, and upper oxidising flames: the former of which, situated at γ in the outer margin of the

FIG. 43.

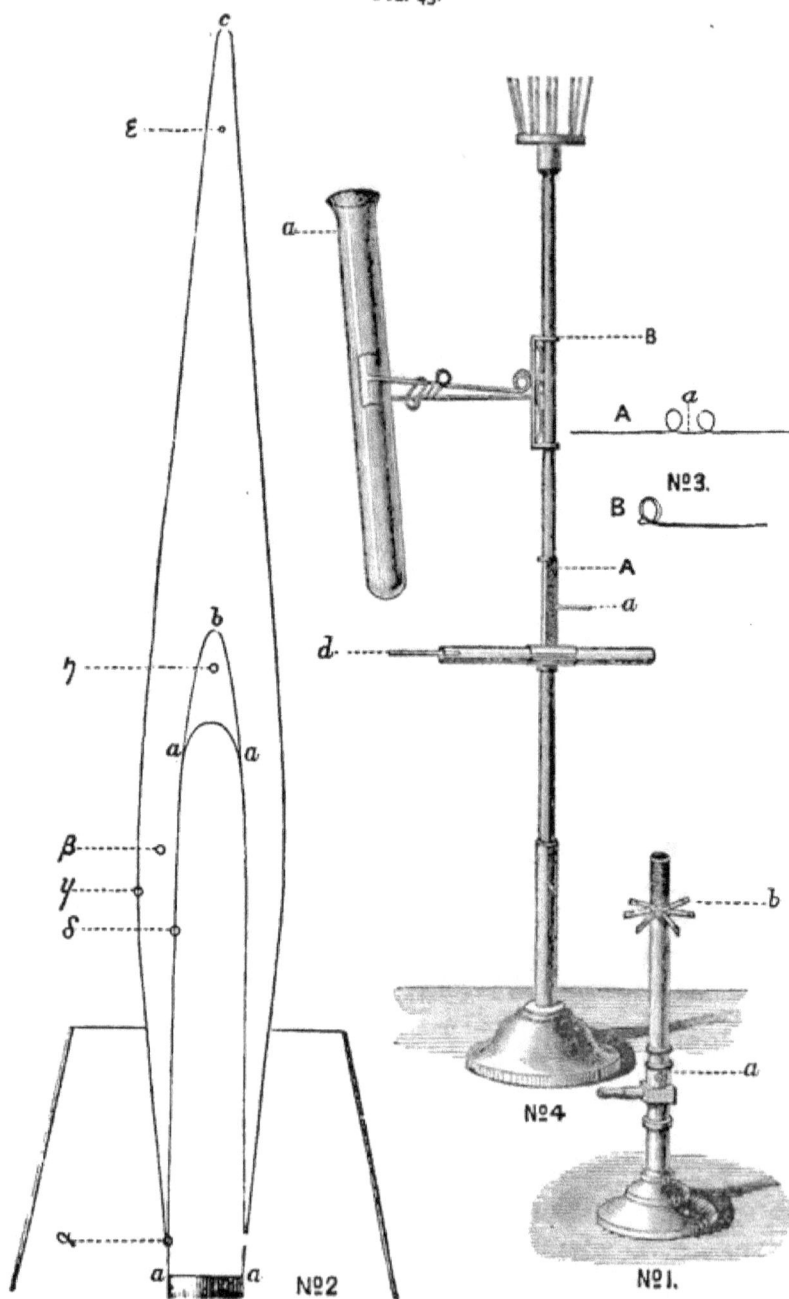

zone of fusion, possesses a higher temperature than the latter, which is found at ϵ, the highest point of the non-luminous flame when the draught-holes are wide open.

5. and 6. The lower, and upper reduction flames: the former is at δ, close to the dark central zone : owing to the presence of atmospheric oxygen the reducing powers of this flame are not so great as those of the upper reducing flame, which is situated at η, just over the dark zone, and is formed by lessening the supply of air so as to produce a considerable amount of finely divided carbon or of dense hydrocarbons, but not sufficient to form a sooty deposit on a test-tube full of cold water held, for a moment, at this point.

In examining substances by means of these flame reactions, the appliances must all be on the smallest possible scale, otherwise so much heat will be carried off by them as to reduce the temperature of the flame to a point at which the proper reaction can no longer be obtained.

Many substances to be tested are held in the flame by means of a little piece of platinum wire ; this wire must be so thin that one decimeter of it does not weigh more than 34 mgm. If two little loops be made on the wire, as shown in fig. 43, No. 3, A, these may then be so brought together, by turning the wire round at the point a, as to form a little catch in which the substance to be tested rests. (See No. 3, B.) Such a platinum loop is also very useful for holding decrepitating substances, which should first be ground to fine powder with the elastic blade of a small knife (fig. 44, No. 1, p. 97) upon a little porcelain plate, then drawn up on to a square centimetre of moistened filtering paper, which, held in the platinum wire loop, can be burned ; the sample is thus obtained adhering to the wire.

Certain substances cannot be held in the flame on the platinum wire, as they either act upon the metal, or do not adhere to it when moistened; these are supported on a piece of asbestos about one-fourth the thickness of an ordinary lucifer-match.

In fig. 43, No. 4, is shown an arrangement by means of which substances to be tested may be held in the flame for a length of time. A small glass tube is held by the arm, *a*, attached to the carrier, A ; into this tube a little piece of platinum wire is fused. The carrier, A, is moveable, so that the substance held in the loop of the wire may be moved to any part of the flame, and held there while the phenomena which occur are noted.

The arm, *d*, also attached to the carrier, A, serves to support threads of asbestos, which are often used instead of platinum.

Another carrier, B, carries an arm which may clasp a small test-tube containing any substance requiring to be held for a length of time in the flame.

From the phenomena noticed when substances are heated we may often learn much concerning them. The following points are more especially to be observed :—

(1) *Emission of Light.*—Heated on the platinum wire in the hottest part of the flame, different substances emit different degrees of light. If the sample and the platinum wire appear equally luminous, the substance is said to be of mean emissive power; if the wire appears more luminous than the substance, the emissive power of the latter is said to be low, while this same quality is called strong or high if the light emitted by the heated substance be more intense than that coming from the platinum wire.

(2) *The melting point* of many substances may be pretty accurately determined by holding them on the platinum wire, in different parts of the flame, and observing the tints assumed by the wire. The following six different temperatures may thus be obtained : (*a*) below a red heat ; (*b*) commencing red heat ; (*c*) red heat ; (*d*) commencing white heat ; (*e*) white heat ; (*f*) strong white heat.

(3) *The colour imparted to the flame* by volatilised substances may be judged of by holding the sample in the upper

reducing flame, when the colour appears in the upper oxidising flame.

If we wish to discriminate between substances which impart various colours to flame, we may often do so by bringing the mixture into the lowest and coldest portion of the flame : the most volatile substance is the first to betray its presence, by imparting a momentary colour to the flame; this is followed by the next volatile, and so on.

But perhaps the most important point you have to notice is the behaviour of substances when heated in the oxidising and reducing flames.

I. REDUCTION OF SUBSTANCES.—Very many compounds when exposed to the influence of the reducing flame give up the oxygen, sulphur, etc., which they contain, the metals being thereby obtained in the pure state; to these various tests can then very easily be applied.

Bring a crystal of common washing soda ($Na_2CO_3 +$ $10H_2O$) to the side of the flame, and when the edge begins to melt, smear a common wooden lucifer-match, with the head broken off, with the fused salt. By gently heating the smeared match in the flame you obtain a mass of carbonised wood intermixed with sodium carbonate. A small quantity of the substance to be tested is now to be placed on the palm of the hand, and well mixed, by means of the small knife, with a little fused soda crystals. A portion of this mixture you must now cautiously bring on to the end of the prepared match. This is, perhaps, the most difficult part of the process, as in attempting to make the substance adhere to the match, the carbonised mass is very apt to break off. Heat the match with the substance on it in the lower reducing flame, situated at δ; the soda effervesces; after a few moments withdraw the match, break off the head into a small agate mortar, press it gently with the pestle, adding a little water; the lighter particles of carbon float upon the water, the soda is dissolved, and the heavy reduced metal is left, on pouring off the water, at the bottom of the mortar. After a few washings the metal is obtained almost perfectly pure. You must now examine the small bead of metal. To do this, remove it to a little piece of glass (a piece of a broken flask is the best), and gently dry it over the flame. The colour and general appearance of the small metallic bead will give you some clue to its nature. Copper may be at once recognised by its red colour; lead, by its bluish-grey colour and its malleability; bismuth and antimony, by their brittleness; iron, by the action of a magnetised steel blade upon it; and so on. If the metal is suspected to be iron, the small knife-blade is drawn once or twice over a magnet, and then brought near to the spicules: if these are attracted by the blade, the metal, you may almost certainly conclude, is iron. Nickel and cobalt are also magnetic, but not to so great a degree as iron.

The metal is further to be tested by being brought into solution and having several *wet tests* applied to it. Let us take an actual example.

Treat a very small quantity of copper sulphate as directed above. From the red colour of the metal obtained you conclude that it is copper; you must, however, make quite sure of this. For this purpose the reduced metal placed on the small glass plate is dissolved by gently warming it with one drop of dilute nitric acid; the excess of acid is driven off by blowing on the surface of the warm glass; a drop of water is added, and the corner of a small piece of filtering paper being dipped into the solution, a very little of it is sucked up by the paper. On now bringing a drop of potassium ferrocyanide (a reagent used as a test for copper) on to the paper, a brown stain of copper ferrocyanide instantly makes its appearance.

Sometimes it is preferable to use small capillary tubes in the application of the wet tests. To form these, heat a piece of wide thin glass tubing, and when soft draw it out slowly, turning it all the while. The thin part of the tube (see fig. 44, No. 4) is now separated from the rest, and cut into lengths of about 6 cm. One of these small capillary tubes is dipped into the solution to be tested; a very small quantity of the solution rises a little way in the tube, from which, by gentle blowing into the other end of the tube, it is expelled on to a small piece of glass; a drop of the test reagent is also brought on to the glass in a precisely similar manner, and after slightly stirring the mixture, a drop of it is again brought into a third capillary tube, when any precipitate or change of colour in the liquid is easily detected.

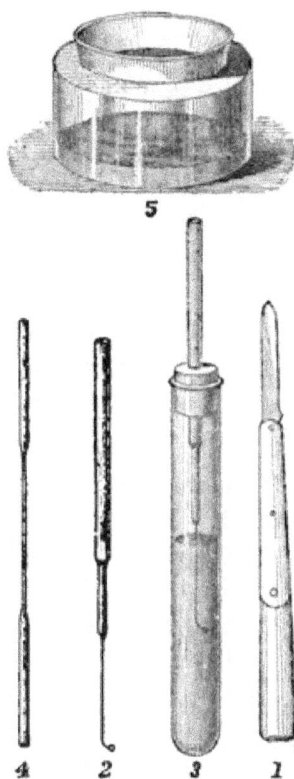

FIG. 44.

Mix a small quantity of ferrous sulphate with the fused sodium carbonate, bring the mixture upon the charred match, and hold it in the upper reducing area (η) for a minute or so. Break off the charred mass containing the reduced metal into a small mortar in which are a few drops of water, crush it gently with the pestle, and insert the magnetised blade of a knife. The spicules of iron will attach themselves to the blade ; they may be washed, without being detached, by allowing a few drops of water to flow along the knife blade, dried, by holding the blade above the flame for an instant, wiped off on to a piece of filter paper, and dissolved in a drop of hydrochloric acid. Dry the paper carefully over the lamp, and moisten the yellow spot with a drop or two of a solution of potassium ferrocyanide : it will be seen to acquire a deep-blue tint owing to the formation of *Prussian blue.*

Treat a little more of the ferrous sulphate on the carbonised splinter with soda in the reducing flame, break off the head of the match on to a silver coin, and moisten it with a drop of water ; in a few moments the coin will be stained brownish-black. This stain tells you that sulphur was present in the substance tested. By the combined action of soda, charcoal, and the reducing flame on the copper sulphate, soluble sodium sulphide has been formed, which, acting upon the silver, has given rise to a film of black silver sulphide. Almost all sulphates may be thus reduced, and the sulphur recognised.

You see how very delicate these reactions in the reducing flame are, and how they may be made the means of detecting almost infinitesimal amounts of many substances.

As you learn the various wet tests for the metals, you will be able yourself to apply them to the beads obtained in your preliminary flame reactions.

But let us now glance at another application of this reducing flame : many of the metals may be obtained, through its influence, in the form of *films.*

The following experiments are only applicable to those elements which are volatile at the temperature of the reducing flame.

The principal of these are antimony, arsenic, bismuth, mercury, and cadmium.

Bring a small piece of arsenious oxide on to a thread of asbestos, by slightly wetting the asbestos between the lips and then touching the oxide with it. Select one or two porcelain basins, 10 to 20 cm. in diameter, and into one pour a little cold water, and hold it close above the asbestos in the upper reducing flame. The arsenious oxide is thus deoxidised, a film of metallic arsenic being deposited on the cold surface of the basin. To this film many tests can now be applied, but these will more properly be performed when you have gone through the wet tests for arsenic. Meanwhile you must note the appearance of the arsenical film or mirror.

Repeat the above experiment, using an antimony compound (antimony trichloride will do) instead of arsenious oxide, and compare the film thus obtained with that given by the arsenic. You at once notice the sooty-black *velvety* appearance of the antimony mirror. From the appearance of the films you may often be able to tell by what metal they are caused.

Volatilise a small quantity of corrosive sublimate ($HgCl_2$) in the manner just described. You thus obtain the metallic film of mercury; you see how it spreads over the entire under-surface of the porcelain dish, and how unlike it is in this respect to either antimony or arsenic.

Although we do not as yet proceed to apply any ' wet tests ' to these metallic films, you will nevertheless perceive the service which may be rendered to the chemist in his qualitative researches by these easily-obtained mirrors.

Let us now look at the results obtainable by the oxidising flame.

Repeat the experiments which you have just performed, but hold the asbestos in the upper oxidising flame ; you will thus obtain films of the oxides of the metals upon the porcelain, instead of the metals themselves.

Obtain thus an oxide film of arsenic, using a small quantity of arsenious oxide. Notice the peculiar lavender tint given to the flame by the arsenical fumes, and remark also the white smoke which rises from the flame, caused by the volatilised oxide of arsenic. On the porcelain you obtain an almost imperceptible film. This, on the application of proper tests, is very easily recognised to be arsenious oxide. Similar oxide films are obtainable from many other metals.

From these oxide films others are obtained which are very characteristic, and often of great use in helping us to draw conclusions as to what substances we are dealing with.

(1) *Iodide Films.*—To obtain these, the dish on which we

have formed the oxide film is, after being breathed on, placed upon the wide-mouthed well-stoppered glass (fig. 44, No. 5) which contains fuming hydriodic acid and phosphorous acid, obtained by the gradual deliquescence of phosphorus tri-iodide. (If this mixture should cease to fume, the addition of a little phosphoric anhydride renders it again fuming.)

(2) *Sulphide Films,* are most easily obtained from the iodide films by blowing a stream of strong ammonium sul-

FIG. 45.

phide over these, and removing excess of the ammonium sulphide by *gently* warming the porcelain basin.

The ammonium sulphide is contained in a small wash bottle, the exit-tube of which is cut short so as not to dip into the liquid, while the blowing tube passes nearly to the bottom of the flask. (See fig. 45.)

The reduction films should be deposited on a glass tube instead of on porcelain when a considerable quantity of the film is required for examination. In this case a tube similar to that represented in fig. 43, No. 4 *a*, is half filled with water, and it is so arranged as to catch and condense the metallic fumes arising from the substance, which, supported on an asbestos thread by the holder *d* (see fig. 43), is held in the upper reduction flame. This arrangement may be left for some time, until a sufficiently large film has accumulated on the glass. (As the water in the glass soon boils, it is desirable to put a very small piece of marble into it to prevent 'bumping.')

A few elements when heated with certain solutions in the Bunsen flame yield coloured beads which are highly characteristic of the different metals. These beads are generally obtained by bringing a small quantity of the substance to be tested on to one of the platinum wires (see fig. 44, No. 2), heating it for a few moments in the hottest part of the flame, allowing it to cool, and by means of a capillary tube moistening the mass with the solution used as a test, and again heating in the oxidising Bunsen flame. Heat a very small quantity of zinc sulphate on the loop of the platinum wire, moisten the bead when cold with cobaltous nitrate solution, and heat again in the upper oxidising flame; a beautiful green-coloured mass is proof of the presence of zinc. The special application of these flame reactions to the various metals will be described when we speak of these in detail.

When we have a mixture of substances each of which colours the Bunsen flame, it is often possible to discriminate between them by viewing the flame in which the mixture is held through coloured glasses, or through a glass prism containing indigo solution. The colours imparted to the flame by different bodies are distinguished as : (1) *border colours*, produced by the most volatile substances when held in the loop of a platinum wire a very little way outside of the flame, about 2 mm. from the lower part ; (2) *mantle colours*, produced by holding certain substances in the outer part or mantle of the flame ; (3) *flame colours*—these colours are shown when the more difficultly volatile substances are held in the hottest part of the flame.

FIG. 46.

A glass prism (fig. 46), filled with dilute indigo solution, is very useful in observing certain flames. The base of the prism may conveniently measure about 35 mm. and its length 150 mm. The indigo solution is made by dissolving 1 part of indigo in 8 parts of fuming sulphuric acid, diluting with 1,500 to 2,000 parts of water, and filtering. The flame is first viewed through the thinner layers of the liquid, and gradually through thicker and thicker layers, by moving the prism horizontally.

If small quantities of a sodium salt be mixed with a salt of potassium, the violet colour given to the flame by the latter is completely hidden by the intense yellow of the

sodium flame; no sodium rays can pass through the indigo
solution, which, however, freely transmits the potassium
rays; if therefore the mixed flame be viewed through the
indigo prism, it appears at first of a sky-blue colour, changing, as it is viewed through thicker layers of the solution,
to violet, and lastly appearing crimson-red.

Other applications of these flame tests with coloured
glasses will be noticed under the special tests for the various
metals.

B. *Use of the Spectroscope.*

Of late years the spectroscope has become of general use
in qualitative analysis.

The indications afforded by this instrument are of such
exceeding delicacy that by its means $\frac{1}{3000000}$ part of a
milligram of sodium can be detected with accuracy.

The spectroscope is shown in fig. 47. It consists of a
flint glass prism, A, having a refracting angle of 60°, resting on
a brass plate which again is fixed on an iron or brass support,
B. The brass plate carries also the tube C. The tubes D
and E are supported by arms, so as to be moveable in a hori-
zontal plane about the axis of the support B.

On the end of the collimator tube, C, nearest the prism,
is placed a lens, the other end being closed by a plate in
which there is a vertical slit. The end of the tube E nearest
the prism is also furnished with a lens; at the other end
is a reduced photographic millimetre scale, which can be
seen through the telescope, D.

It is generally necessary to obtain two spectra in the
field of the telescope at the same time, so that we may compare the positions of the lines in each.

The arrangement of the vertical slit in the collimator
tube, C, enables us to do this. An enlarged drawing of this
arrangement is shown in fig. 48, No. 1. The upper part, *a b*,
of the slit is open; the under part is covered by the small
glass prism, *c*. Rays of light falling directly on the upper

part of the slit pass therefore straight into the tube *c*, as shown in fig. 48, No. 2, p. 104, where A represents the source of light, and the dotted line A *b* the passage of the rays into the tube.

If at the same time rays proceed from a source of light situated at B, they will fall on the surface, *c d*, of the small prism, and will be sent, by total reflection, in the direction *c f b* through the lower part of the slit, also into the colli-

FIG. 47.

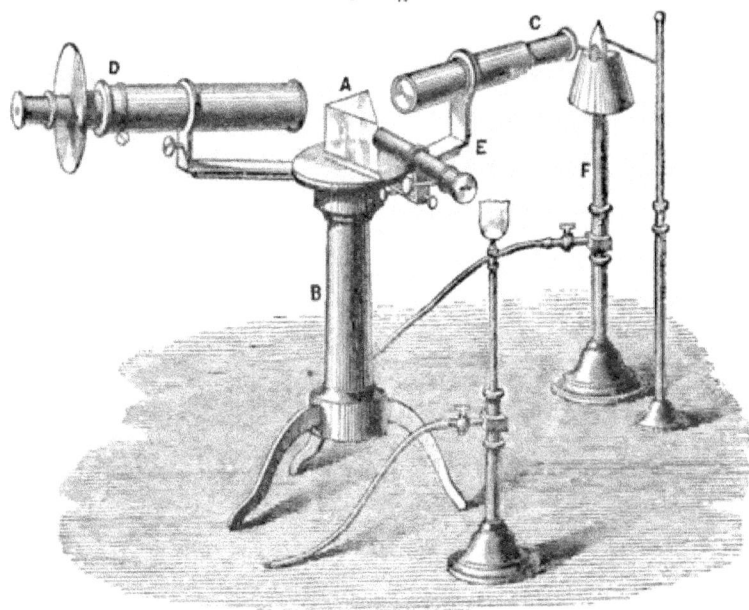

mator tube. By this means we obtain two spectra, one directly over the other, and the coincidence of the lines in each is thus easily observed.

To arrange the instrument, the telescope, D, is so adjusted that a distant object is clearly seen by it ; it is then fixed in its place, and its axis brought into a straight line with that of C, the slit in the latter tube being drawn out, and the telescope adjusted so that the middle of the slit is seen in the middle of the field of view.

The prism is now fixed in its place, and secured by a spring clamp for the purpose. On directing the axis of c towards the flame of a candle, and moving the telescope through a certain angle, the spectrum of the flame is seen in the lower part of the field; E is now placed so that the image of the scale illuminated by a small gas flame is seen through the telescope D.

By means of a small screw on E the divisions are

FIG. 48

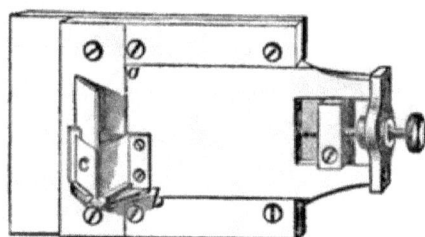

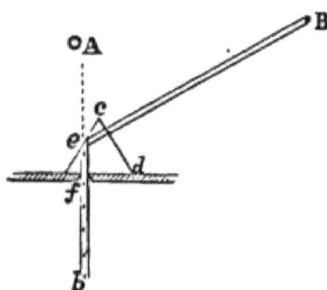

focussed. The breadth of the slit in c can also be regulated by a similar screw. All extraneous rays of light are shut off by throwing a black cloth over the apparatus. The lamp, F, is moved until the mantle of the flame is before the slit. Another lamp is also brought into a proper position, if it be desired to examine two spectra simultaneously.

There is another form of spectroscope now come into very general use, in which the small tube carrying the scale is dispensed with. In this instrument the telescope is moveable upon a graduated circle. The line whose position it is wished to ascertain is brought into the middle of the field of view, and the degrees on the circle are then read off. Several prisms are also often used instead of one; we thus get a higher dispersive power, and consequently a larger spectrum.

In Browning's direct-vision spectroscope (fig. 47 A), which consists of a single tube like a small telescope, the ray of light after entering the slit is refracted and reflected by the prisms in such a manner that it leaves the instrument in the same plane as that in which it entered.

Fig. 47 A.

In using this instrument—which is very compact and serviceable for laboratory use—we look through the tube directly at the object supported in the Bunsen flame. By means of a little device like that described on p. 103, a second spectrum may be sent into the instrument, and the lines of the two spectra compared with one another.

On now bringing a very small quantity of sodium chloride (common salt), supported in the loop of a clean platinum wire, into the mantle of the lamp flame, the yellow line D a is seen strongly marked in the spectrum. This line instantly tells us that sodium is present. Each metal when vapourised, and the vapour heated, exhibits characteristic lines in the spectrum, and by these each is easily recognised. (See Frontispiece.)

It is only when in the state of vapour that the characteristic spectra are afforded by the elements. All solid or liquid substances, when heated, emit light, which, when examined by the spectroscope, is seen at first to consist of red rays only; as the heat increases, other rays are given off, until a point is reached at which, all the luminous rays being emitted, the body is said to be white-hot: such spectra are, therefore, *continuous*, and can afford no indications of the chemical nature of the substance we are examining. The spectra we require are, therefore, discontinuous spectra—spectra made up of rays of different and definite degrees of refrangibility. Such spectra we have produced by analysing the light emitted by glowing gases or vapours. Therefore we can only examine, in the arrangement described above, such substances as are volatile at the highest temperature obtainable by the Bunsen lamp. Other substances are frequently volatilised by means of the electric light or by the spark discharge; but for our purposes it will be sufficient only to test by the spectroscope the more easily volatile bodies.

The application of these tests will be given when we come to speak of the metals in detail.

C. *Use of Preliminary Dry Reactions.*

Many substances when heated in a glass tube exhibit phenomena which enable us to decide with tolerable accuracy what elements are present.

Heat a few crystals of sodium sulphate, reduced to powder, in a small glass tube closed at one end. Drops of water speedily appear on the sides of the tube. These tell you that the salt contained water of crystallisation.

In a similar manner heat a few crystals of potassium chlorate. The salt decrepitates, and then partially fuses. A growing chip of wood on being brought to the mouth of the tube bursts into flame; oxygen is therefore being evolved.

Fig. 49.

Bend a small piece of glass tubing so as to form an obtuse angle (see fig. 49); in the bend place a small quantity of sodium sulphite, and heat this gently. A gas having the smell of burning sulphur is given off.

These experiments may be multiplied greatly; but we mean here merely to point out the great help which preliminary dry reactions yield us in the systematic qualitative examination of an unknown substance. We shall hereafter have opportunities of more fully explaining and developing these reactions.

SECTION II.

SYSTEMATIC QUALITATIVE TESTING.

A. *REACTION OF THE BASES.*

WE shall for the present confine ourselves to the reactions of those metallic bases which are of more common occurrence, viz. :—

Aluminium, antimony, arsenic, barium, bismuth, cadmium, calcium, chromium, cobalt, copper, iron, lead, magnesium, manganese, mercury, nickel, potassium, silver, sodium, strontium, tin, and zinc.

The above-mentioned metallic bases may be divided into a comparatively small number of groups, in accordance with their deportment towards certain reagents, called *group* reagents.

Each group reagent precipitates, under stated conditions, a certain number of metals, and these metals only; and the addition of one of these reagents, if made in proper order, does not hinder the action of the others; that is to say, if group No. 1 be precipitated, and the precipitate removed by filtration, group No. 2 remains in solution, and is thrown down on the addition to the filtrate of the appropriate group reagent.

In analysing a solution containing bases, it is necessary, in the first place, to divide these bases into groups, and then to examine each group precipitate in detail, and to seek for the special metals present by the application of reagents, which give characteristic reactions with the individual metals.

The group reagents are five in number; their names are given in the order of their application :—

(1) Hydrochloric Acid. (2) Sulphuretted Hydrogen. (3) Ammonia in presence of Ammonium Chloride. (4) Ammonium Sulphide. (5) Ammonium Carbonate.

Group I. comprises all metals whose chlorides are practically insoluble in water, and which are therefore precipitable by hydrochloric acid. These are **silver, lead,** and **mercury** (monad mercury or mercurosum).

Group II. comprises all of the remaining metals whose sulphides are insoluble in dilute acids, and which are, therefore, precipitable by sulphuretted hydrogen from acid solutions. These are **mercury** (dyad mercury or mercuricum), **lead, bismuth, copper, cadmium, arsenic, antimony,** and **tin.**

Group III. comprises all of the remaining metals whose hydrates are insoluble in an ammoniacal solution, and which are, therefore, precipitable by ammonia. These are **iron, aluminium,** and **chromium** (manganese is partially precipitated with this group).

Group IV. comprises all of the remaining metals whose sulphides are insoluble in an alkaline solution, and which are, therefore, precipitable by ammonium sulphide. These are **nickel, cobalt, zinc,** and **manganese.**

Group V. comprises all of the remaining metals whose carbonates are insoluble in an alkaline liquid, and which are, therefore, precipitable by ammonium carbonate. These are **barium, strontium,** and **calcium.**

The remaining bases are not classified in groups.

SPECIAL REACTIONS FOR MEMBERS OF GROUP I.

I. SILVER.—Silver nitrate ($AgNO_3$) is used for the tests.

1. *Dry Reactions.*—On charcoal splinter (see p. 94), silver gives a white metallic bead, easily soluble in nitric acid; the solution gives a white precipitate with hydrochloric acid.

2. *Wet Reactions.*—Dissolve a few of the crystals in water, and test the solution as follows :—

(*a*) Hydrochloric acid (HCl) throws down a white curdy precipitate of silver chloride (AgCl); this precipitate is insoluble in hot water, but dissolves in ammonia. On the addition of nitric acid to the ammoniacal solution, silver chloride is reprecipitated.*

(*b*) Sulphuretted hydrogen (H_2S) produces a black precipitate of silver sulphide (Ag_2S).

(*c*) Potassium chromate (K_2CrO_4) throws down dark-red silver chromate (Ag_2CrO_4), soluble in hot nitric acid.

II. MERCURY.—Mercurous nitrate ($HgNO_3$) is used for the reactions.

1. *Dry Reactions.*—On asbestos thread in reducing flame mercury gives a lustrous metallic film spreading over the basin. By breathing on this film, holding it over a dish containing bromine until the black film disappears, and then forming an iodide film, very characteristic carmine-coloured mercuric iodide (HgI_2) is produced.

2. *Wet Reactions.*—Dissolve a few of the crystals in water, and apply the following tests :—

(*a*) *Hydrochloric Acid.*—A white precipitate of mercurous chloride ($HgCl$) is thrown down, insoluble in hot water, insoluble in ammonia, but blackened by this reagent, owing to the formation of a di-mercurous ammonium-chloride (NH_2Hg_2Cl) :

$$2HgCl + 2NH_4HO = NH_2Hg_2Cl + NH_4Cl + 2H_2O.$$

(*b*) *Sulphuretted Hydrogen* produces a black precipitate of mercurous sulphide (Hg_2S).

(*c*) *Stannous Chloride* ($SnCl_2$) throws down a grey precipitate of metallic mercury.

(*d*) Put a slip of clean metallic copper into a little of the solution; a grey stain (metallic mercury) forms on the copper. When rubbed, this stain gets brighter; it is volatilised by heat.

---

* In all cases before testing a precipitate as to its solubility in any liquid, it must be collected on a small filter and thoroughly washed ; a small portion is then removed to a test-tube and the required reagent added.

III. LEAD.—Lead nitrate, $Pb(NO_3)_2$, is used.

1. *Dry Reactions.*—On charcoal splinter, lead salts give soft malleable beads, easily dissolved by nitric acid : the solution gives a white precipitate with sulphuric acid.   On asbestos thread in reducing flame, the flame is coloured pale blue, and simultaneously a black or brown metallic film is formed (the colour is dependent on the thickness of the deposit), from which an orange-yellow iodide film, and a brownish-black sulphide film may be obtained.

2. *Wet Reactions.*—Dissolve the remainder of the crystals in water, and test the solution as follows :—

(*a*) Hydrochloric acid is added ; a white precipitate of lead chloride ($PbCl_2$) forms,* soluble in boiling water (from which solution needle-shaped crystals of the salt are precipitated on cooling), insoluble in ammonia.

(*b*) Sulphuric acid (H_2SO_4) throws down a heavy white precipitate of lead sulphate ($PbSO_4$), insoluble in acids, but soluble in caustic soda or in an ammoniacal solution of tartaric acid ($H_2C_4H_4O_6$).

(*c*) Potassium iodide (KI) precipitates yellow lead iodide (PbI_2), which is soluble in boiling water ; on cooling, brilliant golden scales of the salt are precipitated.

On examining these reactions, we notice that lead chloride is soluble in hot water, whereas the other chlorides of the group are insoluble ; further, that silver chloride is dissolved by ammonia, mercurous chloride remaining undissolved, but being blackened by this reagent. A method of separation is thus easily framed.

Separation of Group I.—To a solution containing the three metals add hydrochloric acid, collect the ensuing precipitate on a small filter, wash it once or twice with *cold* water, remove it to a test-tube, and boil it with water ; filter while hot, and to the filtrate add sulphuric acid ; a white

---

* As lead chloride is sparingly soluble in cold water, traces of the metal will be present in the filtrate. and may be detected by the methods given under the next group.

precipitate indicates lead (III. *b.*). Warm a little strong ammonia solution, and pour it over that portion of the group precipitate which is insoluble in water; it is blackened, showing the presence of mercury (II. *a*). Acidulate the ammoniacal filtrate with nitric acid · a white precipitate tells you that silver is present (I. *a*).

SPECIAL REACTIONS FOR MEMBERS OF GROUP II.

I. MERCURY.*—Mercuric nitrate, $Hg(NO_3)_2$, is used.

1. *Dry Reactions.*—Same as those for mercurous compounds. (See p. 108.)

2. *Wet Reactions.*—Dissolve the salt in water and apply the following tests :—

(*a*) Sulphuretted hydrogen at first produces a white precipitate ($HgS.HgNO_3$), which on the further addition of the reagent goes through a gradation of colours, black mercuric sulphide (HgS) being the last product. This precipitate is insoluble in ammonium sulphide, in hydrochloric or nitric acids, but is dissolved by boiling aqua regia.

(*b*) Stannous chloride produces at first a greyish precipitate of mercurous chloride, which changes, on addition of excess of the precipitant, to grey metallic mercury.

(*c*) Copper foil—same as with mercurous salts (p. 109, II. *d*).

II. BISMUTH.—Bismuth chloride ($BiCl_3$) is used.

1. *Dry Reactions.*—On asbestos thread in reducing flame, a black metallic film, thin part brown, difficultly soluble in nitric acid. In oxidising flame, a yellowish-white oxide film. On charcoal splinter, yellow shining spiculæ, soluble in nitric acid.

2. *Wet Reactions.*—Dissolve in dilute hydrochloric acid, and test as follows :—

(*a*) Sulphuretted hydrogen throws down a brownish-

---

* There are two classes of mercury salts, the type of the first being calomel, $HgCl$; and that of the second corrosive sublimate, $HgCl_2$. Distinct tests are given for each class.

black precipitate of bismuth sulphide (Bi_2S_3), insoluble in ammonium sulphide, but soluble in hot nitric acid.

(*b*) Ammonia throws down white bismuthous oxide (Bi_2O_3), which, collected on a small filter, washed, and dissolved in a very little hydrochloric acid (excess of acid being driven off by heat), again forms a solution of bismuth trichloride ($BiCl_3$), from which

(*c*) Water precipitates white insoluble bismuth oxychloride (*bismuthyl chloride* BiOCl), insoluble in tartaric acid. *Compare VI. (c).*

III. Copper.—Cupric nitrate, $Cu(NO_3)_2$, is used.

1. *Dry Reactions.*—On charcoal splinter, red metallic bead, soluble in nitric acid ; the solution gives a mahogany colouration with potassium ferrocyanide (best seen by sucking up a little of the liquid on to filter paper). With borax bead copper gives a blue glass; if a trace of a tin salt is added, and the bead heated in the lower reducing flame, the colour changes to red-brown, owing to the formation of cuprous oxide (Cu_2O). If this bead be alternately slowly oxidised and reduced, a ruby-red colour is obtained.

Small quantities of copper may be detected by bringing the substance to be tested, along with a little pure silver chloride made into a paste with water, supported on a thin iron wire, into the reducing flame, when a fine blue colour is imparted to the flame. (Gericke.)

2. *Wet Reactions.*—Dissolve the remaining crystals in water, and apply the following tests :—

(*a*) Sulphuretted hydrogen precipitates black cupric sulphide (CuS), nearly insoluble in ammonium sulphide and in hot dilute sulphuric acid, soluble in hot nitric acid and in potassium cyanide solution.

(*b*) Ammonia, added drop by drop, produced a greenish-blue precipitate $2(CuSO_4.2CuO).5H_2O$ (Reindel), which disappears in excess of ammonia, a deep-blue liquid containing a double salt of ammonium and copper, $Cu(NO_3)_2 \ 4NH_3$, being formed.

(*c*) Potassium ferrocyanide (K_4FeCy_6) produces a brown precipitate, or in very dilute solution a brown colouration, of copper ferrocyanide (Cu_2FeCy_6)

IV. CADMIUM.—Metallic cadmium is dissolved in nitric acid, the solution evaporated to dryness, and the dry residue tested as follows :—

1. *Dry Reactions.*—On asbestos thread in reducing flame, a black film, thin part brown, instantly soluble in dilute nitric acid. Sulphide film lemon-coloured. On charcoal splinter, partly reduced, giving white ductile beads. Oxide film brown-black, passing through changes of colour till the outer part is reached, which is nearly white.

2. *Wet Reactions.*—Dissolve part of the residue—cadmium nitrate, $Cd(NO_3)_2$—in water, and apply the following tests :—

(*a*) Sulphuretted hydrogen throws down yellow cadmium sulphide (CdS), insoluble in ammonium sulphide, also in potassium cyanide, but dissolved by hot dilute sulphuric acid, also by hot nitric acid.

(*b*) Ammonia precipitates white cadmium hydrate $Cd(OH)_2$, soluble in excess of the precipitant.

V. ARSENIC.—Arsenious oxide (As_2O_3) is used.

1. *Dry Reactions.*—On asbestos thread in reducing flame, black film, thin part brown, soluble in sodium hypochlorite, insoluble in dilute nitric acid. Sulphide film lemon-yellow, disappears when excess of ammonium sulphide is used, but reappears on breathing on the film. On asbestos thread in oxidising flame, white film with white fumes of arsenious oxide and pale lavender-blue flame. If this film be touched with a drop of ammonio-silver nitrate, it gives a yellow colour.

Heated in a bent tube (see p. 106), arsenious oxide sublimes and condenses on the cold part of the tube in octahedral crystals.

Heated in a tube closed at one end along with a fragment of charcoal (fig. 50), arsenious oxide is reduced, a
mirror of metallic arsenic being formed on the cold
part of the tube.

Fig. 50.

Heated in a closed tube with sodium acetate,
cacodyl, $As(CH_3)_2$, is formed, recognisable by its
strong garlic odour. This substance is extremely
poisonous. Arsenic in presence of copper may be
readily recognised by bringing the substance to be
tested into the oxidising flame, and noticing the
peculiar greyish blue flame ; then treating the residue
with silver chloride, as directed on p. 112.

2. *Wet Reactions.*—Boil a little of the arsenious oxide
in water, and to this solution apply the following tests :—

(*a*) Sulphuretted hydrogen precipitates yellow arsenious
sulphide (As_2S_3), insoluble in concentrated boiling hydrochloric acid, soluble in ammonium carbonate, also in
ammonium sulphide, with formation of a sulphur salt, thus :

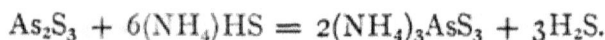

$$As_2S_3 + 6(NH_4)HS = 2(NH_4)_3AsS_3 + 3H_2S.$$

On the addition of acid this salt is decomposed, arsenious
sulphide being reprecipitated, thus :

$$2(NH_4)_3AsS_3 + 6HCl = As_2S_3 + 6NH_4Cl + 3H_2S.$$

(*b*) Ammonio-silver nitrate (*reagent list*) precipitates
yellow silver arsenite (Ag_3AsO_3), soluble in excess of ammonia, also in nitric acid.

(*c*) Ammonio-cupric sulphate (*reagent list*) precipitates
pale green copper-hydrogen-arsenite ($CuHAsO_3$).

Boil a small quantity of arsenious oxide in nitric acid ;
evaporate off the greater part of the acid, dilute with water,
and test the solution of arsenic acid (H_3AsO_4), thus obtained,
as follows :—

(*d*) After neutralisation with ammonia, silver nitrate
throws down brick-red silver arsenate (Ag_3AsO_4), soluble in
ammonia and nitric acid.

(*e*) Add excess of ammonia, then ammonium chloride

and magnesium sulphate, a white precipitate of ammonium-magnesium arsenate, $(NH_4)MgAsO_4$, is produced.

(*f*) Add a strong solution of caustic potash and a few pieces of metallic zinc, cover the mouth of the tube with paper moistened with silver nitrate, and very gently warm the tube. Arsenuretted hydrogen (AsH_3) is evolved, which is decomposed by the silver nitrate, a bluish-black stain $(AsAg_3)$ being produced.

(*g*) Add a few drops of pure hydrochloric acid and a strip of bright copper foil, and boil the liquid. A lustrous brownish-black deposit $(Cu_5As_2$, according to Lippert) is produced. The copper is removed, dried by pressing between folds of filtering paper, placed in a clean dry test-tube, and gently heated. The arsenic is volatilised and oxidised, and condenses on the colder part of the tube as arsenious oxide. The sublimate is dissolved in water, and tested with ammonio-silver nitrate and ammonio-copper sulphate (vide *supra*).

The hydrochloric acid used in this test must be free from sulphites or sulphurous acid, otherwise cuprous chloride is formed, which interferes with the subsequent reaction for arsenic.

VI. ANTIMONY.—Antimonious chloride $(SbCl_3)$ is used.

1. *Dry Reactions.*—On asbestos thread, in reducing flame, black metallic film, more lustrous and velvet-like than the arsenical film, insoluble in sodium hypochlorite. Sulphide film orange, disappearing when large excess of ammonium sulphide is used, but reappearing on blowing. Oxide film white; touched with ammonio-silver nitrate, and blown on with ammoniacal air, gives a black spot insoluble in ammonia.

On charcoal splinter yields white brittle metallic bead. In glass tube does not sublime.

Flame colouration pale greenish-blue, appearing greener on addition of silver chloride.

Dissolve part of the crystals in dilute hydrochloric acid and apply the following tests :—

(*a*) Sulphuretted hydrogen produces an orange-red precipitate of antimonious sulphide (Sb_2S_3) insoluble in ammonium carbonate, but dissolved by boiling concentrated hydrochloric acid, also by ammonium sulphide. See V. (*a*), p. 114.

(*b*) Ammonia throws down white antimony hydrate, $Sb(OH)_3$.

(*c*) On the addition of water, white antimony oxychloride, *powder of algaroth* (SbOCl), is precipitated. This substance dissolves in tartaric acid. Compare II. (*c*), p. 112.

(*d*) A rod of zinc is immersed in a part of the solution contained in a platinum basin. Metallic antimony is deposited as a black stain on the platinum, and is found to be insoluble in cold hydrochloric acid, but to be immediately dissolved by warm nitric acid.

(*e*) Treated as directed in V. (*f*), p. 115, antimony compounds do not evolve antimonuretted hydrogen.

VII. TIN.—Stannous chloride ($SnCl_2$) is used—

1. *Dry Reactions.*—On charcoal splinter tin gives a white ductile bead, soluble in nitric acid, the solution giving a white precipitate with mercuric chloride. If a borax bead be coloured faintly blue by a copper salt, and a trace of a tin salt added, the colour of the bead changes to red when heated in the lower reducing flame.

2. *Wet Reactions.*—Dissolve some of the stannous chloride in dilute hydrochloric acid, and apply the following tests :—

(*a*) Sulphuretted hydrogen throws down dark-brown stannous sulphide (SnS); this precipitate is soluble in yellow (not in colourless) ammonium sulphide ; from this solution hydrochloric acid precipitates yellow stannic sulphide (SnS_2).

(*b*) Mercuric chloride throws down mercurous chloride; but upon addition of a further quantity of the reagent, grey metallic mercury is formed.

(*c*) Zinc, under the same conditions as in VI. (*d*), precipitates metallic tin, chiefly as a grey spongy mass on the zinc.

Dissolve a little tin in nitric acid, and boil down the solution ; stannic oxide (SnO_2) is formed ; when this is of a pure white colour, pour off the acid, wash with water, boil with concentrated hydrochloric acid, dilute the solution, and apply the following tests :—

(*d*) Sulphuretted hydrogen gives a yellow precipitate of stannic sulphide, soluble in boiling concentrated hydrochloric acid, soluble in ammonium sulphide, insoluble in ammonium carbonate.

(*e*) Boiled with a large excess of water, stannic hydrate, $SnO(HO)_2$, is precipitated.

(*f*) Zinc gives the same reactions as with stannous salts.

Separation of Group II.—All the members of this group are precipitated by sulphuretted hydrogen ; but the sulphides of the three last-mentioned metals are soluble in ammonium sulphide, while those of the other metals are insoluble. The group is, therefore, divided into two sub-groups.

Further, sulphide of mercury is insoluble in nitric acid, whereas the other sulphides of Division I. are soluble.

Bismuth is precipitated by ammonia in excess, copper and cadmium are not.

By taking advantage of these and a few other properties of the members of the group, a method of separation may be readily framed.

After washing the group precipitate produced by adding sulphuretted hydrogen to a slightly acid solution of the members of the group, transfer it to a test-tube, and warm it gently with a little yellow ammonium sulphide and a small quantity of water ; filter, preserve the filtrate (labelling it *Div. II.*). Wash the precipitate *thoroughly* with hot water, transfer it to a test-tube, and heat it with dilute nitric acid, filter from undissolved mercuric sulphide, wash thoroughly, and set aside the filtrate (labelled *Div. I., Cu,* &c.); dissolve the mercuric sulphide in aqua regia, boil off excess of acid, dilute with water, and test with stannous chloride (p. 109, I. (*b*)).

To filtrate *Div. I.*, *Cu*, &c., add ammonia in excess ; filter off the ensuing precipitate of bismuth hydrate (test it further by II. (*b*), p. 112), acidify the filtrate—which is blue, showing the presence of copper—with hydrochloric acid, and add sulphuretted hydrogen ; collect the precipitate of copper and cadmium sulphides, and, *after thorough washing*, remove it to a test-tube, and heat it with dilute sulphuric acid ; filter from undissolved cupric sulphide, and test the filtrate with sulphuretted hydrogen ; a yellow precipitate indicates the presence of cadmium. Confirm the presence of copper by dissolving the portion insoluble in sulphuric acid in nitric acid, and adding potassium ferrocyanide. (See p. 113.)

The solution labelled *Div. II.* is neutralised with hydrochloric acid, the ensuing precipitate collected, well washed, and very gently heated with a strong solution of ammonium carbonate ; the insoluble part is removed by filtration, and the filtrate acidified with hydrochloric acid ; yellow arsenious sulphide is thrown down. This may be further tested by dry tests (see V. p. 113). The portion insoluble in ammonium carbonate is dried, mixed with three parts of fused ammonium nitrate, and projected, in small quantities, into a porcelain crucible containing two parts of the same salt in a state of fusion. When fumes cease to be evolved the fusion is stopped, and the mass, which now contains stannic oxide and ammonic antimoniate, when cool, treated with a saturated solution of tartaric acid ; after filtration, hydrochloric acid and sulphuretted hydrogen are added to the liquid, when orange-red antimonious sulphide is precipitated ; the portion insoluble in tartaric acid is tested on the charcoal splinter for tin. (See VII. p. 116.)

SPECIAL REACTIONS FOR THE MEMBERS OF GROUP III.

I. IRON.—(*a*) As ferrous salts. Ferrous sulphate ($FeSO_4$) is used.

1. *Dry Reactions.*—On charcoal splinter in reducing flame yields minute spiculæ, which are magnetic, soluble in hydrochloric acid, the solution giving a deep blue precipitate with potassium ferrocyanide. With borax bead in oxidising flame gives a yellowish brown, in reducing flame, a bottle-green glass.

2. *Wet Reactions.*—Dissolve a few of the crystals in water, and apply the following tests : —

(*a*) Ammonia precipitates white ferrous hydrate, $Fe(HO)_2$, which absorbs oxygen from the air, changing its colour to green, and eventually being converted into reddish-brown ferric hydrate, $Fe(HO)_3$.

(*b*) Ammonium sulphide throws down, after a little time if the solution be dilute, a black precipitate of ferrous sulphide (FeS), soluble in hydrochloric acid with evolution of sulphuretted hydrogen.

(*c*) Potassium ferrocyanide (K_4FeCy_6) produces a blue precipitate—Prussian blue (Fe_7Cy_{18})—insoluble in acids, but decomposed by alkalies.

(*d*) Potassium ferricyanide (K_3FeCy_6) throws down a blue precipitate—Turnbull's blue (Fe_5Cy_{12}).

Boil a little of the ferrous sulphate solution with a few drops of strong nitric acid until the liquid becomes of a yellow colour ; the iron is thus oxidised to the ferric state. From this solution—

(*e*) Ammonia at once throws down foxy-red ferric hydrate.

(*f*) Sulphuretted hydrogen produces a precipitate of sulphur, while the ferric is reduced to a ferrous salt ; thus with ferric chloride :

$$Fe_2Cl_6 + H_2S = 2FeCl_2 + 2HCl + S.$$

(*g*) Add ammonia until the solution is exactly neutral, and then ammonium acetate, and boil the liquid ; a brownish-red precipitate, consisting of a mixture of basic ferric acetates, forms.

(*h*) Potassium ferrocyanide throws down Prussian blue. (See *c.*)

(*i*) Potassium ferricyanide produces no precipitate, but changes the colour of the solution to reddish brown.

(*k*) Potassium sulphocyanide (KCNS), even in very dilute solutions, produces a blood-red colouration, due to the formation of a soluble iron sulphocyanide, $Fe(CNS)_3$.

(*l*) Barium carbonate suspended in water precipitates ferric hydrate mixed with basic salt, carbon dioxide being evolved.

II. ALUMINIUM.—Alum is used for the tests.

1. *Dry Reactions.*—Compounds of this metal heated on platinum wire in the oxidising flame, moistened with cobaltic nitrate, $\left(CO(NO_3)_2\right)$, and again heated in the flame, give a deep blue-coloured mass.   If the precipitate to be tested be very small, collect it on a minute filter, spread out the filter on a flat surface, cut away that portion on which there is no precipitate, place the filter containing the precipitate between the loops of the platinum wire (see p. 92, fig. 43), gently dry the paper, incinerate it, moisten the ash with a drop of cobaltic nitrate solution from a capillary tube, and again heat in the oxidising flame.

The tests here given fail to detect aluminium when present in such a substance as bread.

A part of the bread in crumbs is ignited in a platinum dish ; when the ash is white, or nearly so, it is warmed with a little pure hydrochloric acid, excess of acid being removed by heat.   After cooling, a little piece of sodium, cut from the inside of a lump, is added, and then, very carefully, distilled water.   When the sodium is completely oxidised the contents of the basin are boiled, filtered, the filtrate acidulated with nitric acid, and ammonium carbonate added.

If no precipitate appears after standing, alumina is absent. If a precipitate forms, it is collected on a very small filter, and tested as directed on p. 120.

2. *Wet Reactions.*

(*a*) Ammonia produces a white precipitate of aluminic hydrate, Al(HO)$_3$.

(*b*) Ammonium sulphide produces the same precipitate.

(*c*) Caustic soda or potash throws down the same precipitate ; it is soluble in an excess of the reagent, but reprecipitated by ammonium chloride.

(*d*) Barium carbonate also precipitates aluminic hydrate mixed with basic salt.

III. CHROMIUM.—Potassium bichromate is used for the tests.

1. *Dry Reactions.*—Heated on platinum foil with potassium nitrate, a yellow mass of potassium chromate (K$_2$CrO$_4$), soluble in water, is formed.

On borax bead in oxidising and reducing flames, gives a green coloured bead.

2. *Wet Reactions.*—Dissolve a few crystals of potassium bichromate in water, add a little hydrochloric acid and alcohol, and boil until the liquid is dark green in colour. This solution now contains chromium chloride :

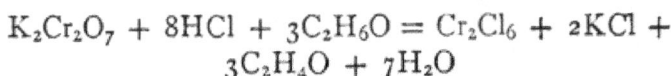

$$K_2Cr_2O_7 + 8HCl + 3C_2H_6O = Cr_2Cl_6 + 2KCl + 3C_2H_4O + 7H_2O$$

(*a*) Ammonia precipitates greyish-green chromic hydrate, Cr(OH)$_3$.

(*b*) Caustic soda or potash produces the same precipitate. It is soluble in an excess of the precipitant, but is reprecipitated on boiling for a few minutes.

Chromium in certain compounds (the chromates) plays a chlorous rather than a basylous part. Its presence in such substances may be detected as follows. Using as example a solution of potassium dichromate in water.

(*a*) Ammonium sulphide throws down a brown precipitate consisting chiefly of chromium chromate, $Cr_2(CrO_4)_3$, the solution now containing potassium chromate, thus—

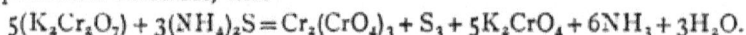

$$5(K_2Cr_2O_7) + 3(NH_4)_2S = Cr_2(CrO_4)_3 + S_3 + 5K_2CrO_4 + 6NH_3 + 3H_2O.$$

If ammonium sulphide be added in excess and the liquid boiled, green chromic hydrate, $Cr(OH)_3$, mixed with sulphur, is precipitated—

$$K_2Cr_2O_7 + 3(NH_4)_2S + H_2O = 2Cr(OH)_3 + S_3 + 2KHO + 6NH_3.$$

(*b*) Lead acetate, $Pb(C_2H_3O_2)_2$ precipitates yellow lead chromate $PbCrO_4$. Soluble in caustic soda.

Manganese belongs to Group IV. ; but if present in a solution it is generally partially precipitated along with the members of this group.

Separation of Group III.—Collect the precipitate produced by addition of ammonium chloride and ammonia, wash it thoroughly, dissolve it in hydrochloric acid, and boil down the solution to a small bulk with addition of a few crystals of potassium chlorate ; allow the liquid to cool, dilute it with water, add caustic soda in excess, and boil ; filter, acidulate the filtrate with hydrochloric acid, and add ammonia : a white precipitate, $Al(HO)_3$, shows the presence of aluminium.

The precipitate by caustic soda is transferred to a platinum dish, dried, and fused with potassium nitrate, the fused mass boiled with water, and the solution filtered from the insoluble portion. To the filtrate acetic acid and lead acetate are added—a yellow precipitate ($PbCrO_4$) indicates the presence of chromium. The insoluble part of the fused mass is dissolved in hydrochloric acid and tested for iron by potassium ferrocyanide (I. (*c*), p. 119).

SPECIAL REACTIONS FOR THE MEMBERS OF GROUP IV.

I. NICKEL.—Nickelous sulphate ($NiSO_4$) is used.

1. *Dry Reactions.*—On charcoal splinter nickel gives white ductile particles, which are slightly magnetic ; dissolved in nitric acid a green-coloured solution is produced, which on paper (p. 96) moistened with soda, exposed to bromine

vapour, and again treated with soda, gives a black stain (Ni_2O_3). On borax bead in oxidising flame, gives a greyish-brown to red-brown glass; in upper reducing flame, grey (from reduced nickel).

2. *Wet Reactions.*—Make a solution of the salt in water, and apply the following tests:—

(*a*) Ammonium sulphide produces a black precipitate of nickelous sulphide (NiS), almost insoluble in cold dilute hydrochloric acid, soluble in aqua regia.

(*b*) Caustic soda or potash precipitates light green nickelous hydrate, $Ni(HO)_2$.

(*c*) Ammonia causes the formation of the same precipitate, soluble in an excess of the precipitant, forming a blue fluid from which nickelous hydrate is reprecipitated by caustic potash.

(*d*) Potassium cyanide* (KCy) throws down yellow nickel cyanide, soluble in an excess of the precipitant, forming a brownish-yellow solution ($2KCy.NiCy_2$). From this solution when boiled with sodium hypochlorite solution (NaClO) black nickelic hydrate, $Ni(HO)_3$, precipitates.

From this same solution, when hot, nickelous hydrate is precipitated on adding a solution of mercuric oxide in mercuric cyanide,† and allowing the mixture to stand for a little time.

II. Cobalt.—Cobaltic nitrate, $Co(NO_3)_2$, is used.

1. *Dry Reactions.*—On charcoal splinter in reducing flame, white ductile particles, slightly magnetic, soluble in nitric acid, giving a reddish-coloured solution. Draw up a little of this solution on to filter paper and dry it; the colour changes to green.

In borax bead in either flame cobalt gives a bright blue glass. When heated in the upper reducing flame for some

---

* The potassium cyanide should be purified from cyanate by fusion in a porcelain crucible with charcoal, and solution in water.

† To prepare this reagent boil mercuric oxide in an aqueous solution of mercuric cyanide, and filter from undissolved oxide.

time the bead becomes grey-coloured, owing to the separation of metallic cobalt.

2. *Wet Reactions.*—Dissolve a little of the salt in water, and apply the following tests:—

(*a*) Ammonium sulphide precipitates black cobaltous sulphide, CoS, insoluble in cold dilute hydrochloric acid, soluble in aqua regia.

(*b*) Caustic soda or potash precipitates a blue basic salt, which on boiling is converted into red hydrate, $Co(HO)_2$.

(*c*) Ammonia produces the same precipitate; it is soluble in an excess of the reagent, yielding a reddish-brown liquid.

(*d*) Potassium cyanide * precipitates light-brown cobaltic cyanide ($CoCy_2$), soluble in an excess of the reagent, the solution containing $2KCy.CoCy_2$, but when boiled this is changed to potassium cobalti-cyanide (K_3CoCy_6), which is not precipitated by sodium hypochlorite or by mercuric oxide in mercuric cyanide.

III. ZINC.—Zinc sulphate, $ZnSO_4 + 7H_2O$, is used.

1. *Dry Reactions.*—On asbestos thread in reducing flame zinc gives a black metallic film, thin part brown.

In oxidising flame yields a white oxide film. Heated on platinum wire, moistened with cobaltic nitrate, and again heated, gives a green-coloured mass. In testing for small quantities of zinc salts, use the method described under aluminium (p. 120).

2. *Wet Reactions.*—Dissolve the salt in water, and apply the tests as follows:—

(*a*) Ammonium sulphide precipitates white zinc sulphide (ZnS), soluble in mineral acids, insoluble in caustic potash or soda, and in acetic acid.

(*b*) Caustic potash or soda throws down white zinc hydrate, $Zn(HO)_2$, soluble in excess of the precipitant; the solution thus obtained is not reprecipitated by ammonium chloride.

* See note p. 123.

IV. MANGANESE.—Manganese chloride ($MnCl_2$) is used.

1. *Dry Reactions.*—On borax bead in the oxidising flame manganese gives an amethyst-coloured glass, which becomes colourless in the reducing flame.

If a manganese salt be heated with sodium carbonate and a little potassium nitrate, a green bead is formed.

2. *Wet Reactions.*—Dissolve a little manganese chloride in water, and apply the following tests to the solution :—

(*a*) Ammonium sulphide precipitates flesh-coloured manganous sulphide (MnS), soluble in acids (including acetic acid).

(*b*) Caustic potash or soda throws down a dirty white precipitate of manganous hydrate, $Mn(HO)_2$, insoluble in an excess of the precipitant. By exposure to the air this hydrate is oxidised and darkens in colour. After such oxidation it is insoluble in ammonium chloride.*

(*c*) Treat a small quantity of the solution of any manganese salt (free from chlorine) with lead dioxide (PbO_2), add nitric acid (free from chlorine), and boil ; the liquid assumes a reddish-purple colour, owing to the formation of permanganic acid ($HMnO_4$).

Separation of Group IV.—To a solution containing all the members of the group add ammonium sulphide in slight excess ; collect the precipitate as rapidly as possible on a small filter, and wash with water containing a little ammonium sulphide, to prevent oxidation of the sulphides. Remove the precipitate to a test-tube, shake it up with dilute hydrochloric acid, filter, to the filtrate add caustic potash in excess ; a brown precipitate, $Mn(HO)_2$, shows the presence of manganese. Again filter and add ammonium sulphide to the filtrate ; white-zinc sulphide (ZnS) is precipitated. That portion of the group precipitate which is insoluble in

---

* This explains the fact that manganese is partially precipitated along with the members of Group III.

hydrochloric acid is dissolved in aqua regia, the acid solution evaporated almost to dryness, diluted, nearly neutralised with sodium carbonate, and potassium cyanide (see note p. 123), added until the precipitate which at first forms is redissolved, the liquid is then boiled, and while still hot a solution of mercuric oxide in mercuric cyanide is added; the pale green precipitate of nickelous hydrate is collected on a filter and tested by means of a borax bead (see I. p. 122), the filtrate is evaporated to dryness, and the residue tested in the borax bead for cobalt (see p. 124).

SPECIAL REACTIONS FOR THE MEMBERS OF GROUP V.

I. BARIUM.—Barium chloride ($BaCl_2$) is used.

1. *Dry Reactions.*—Supported on platinum wire, and heated in the zone of fusion, barium salts impart a green colour to the flame, appearing bluish-green when viewed through the green glass (see p. 101).

Certain barium salts, especially phosphates and silicates, must be moistened with hydrochloric or sulphuric acid before being brought into the flame.

2. *Wet Reactions.*—To a solution of the chloride in water add—

(*a*) Ammonia and ammonium carbonate; a white precipitate of barium carbonate ($BaCO_3$) is thrown down.

(*b*) Calcium sulphate produces an instantaneous turbidity, owing to the formation of insoluble barium sulphate ($BaSO_4$).

(*c*) Sulphuric acid precipitates barium sulphate as a heavy white powder.

(*d*) Ammonium oxalate throws down white barium oxalate (BaC_2O_4), soluble in hydrochloric and nitric acids.

II. STRONTIUM.—Strontium chloride is used.

1. *Dry Reactions.*—Strontium salts impart a red colour to the flame, appearing purple or rose when viewed through the blue glass.

2. *Wet Reactions.*—To an aqueous solution of the chloride apply the following tests :—

(*a*) Ammonia with ammonium carbonate precipitates white strontium carbonate (SrCO$_3$).

(*b*) Calcium sulphate solution produces a slight turbidity, but only after standing for some time (SrSO$_4$).

(*c*) Sulphuric acid precipitates white strontium sulphate.

III. CALCIUM.—Calcium chloride is used.

1. *Dry Reactions.*—Calcium salts impart a yellowish-red colour to the flame, appearing greenish-grey when viewed through the blue glass.

On moistening a mixture of salts of the three metals barium, strontium, and calcium with sulphuric acid, carefully drying it, and bringing it, on a platinum wire, into the zone of fusion, the green colouration due to barium is at once visible. When this has disappeared, and when after moistening with hydrochloric acid no blue-green tint appears on viewing the flame through the green glass, the sample is again moistened with hydrochloric acid, and brought into the flame; while spirting it is examined through the blue glass for calcium (see III. above). After the greenish-grey colour has disappeared, the rose-purple, which shows the presence of a strontium salt, is noticed. (Fresenius.)

2. *Wet Reactions.*—Dissolve the salt in water, and test the solution as follows :—

(*a*) Ammonium carbonate produces a white precipitate of calcium carbonate (CaCO$_3$).

(*b*) Calcium sulphate produces no precipitate.

(*c*) Sulphuric acid precipitates calcium sulphate, but only from strong solutions of calcium salts.

(*d*) Ammonium oxalate produces a white precipitate of calcium oxalate (CaC$_2$O$_4$), soluble in hydrochloric and nitric acids ; insoluble in oxalic and acetic acids.

Separation of Group V.—Collect the precipitate produced by the addition of ammonia, ammonium chloride and

ammonium carbonate, wash it well, and dissolve it in hydrochloric acid ; add ammonium sulphate solution and boil ; filter
off the precipitated barium and strontium sulphates, and test
the filtrate for calcium by means of ammonium oxalate (see
III. (*d*) p. 127). Boil the mixed sulphates of barium and calcium with a solution of two parts of potassium carbonate and
one part of potassium sulphate, filter while hot, wash with hot
water, and treat the precipitate on the filter with hydrochloric acid ; the insoluble portion on the filter is barium
sulphate (confirm by flame test, see I. p. 126). To the
hydrochloric acid solution, calcium sulphate solution is
added ; a white precipitate, forming after a little time, shows
the presence of strontium.

SPECIAL REACTION FOR THE REMAINING METALS.

I. MAGNESIUM.—Magnesium sulphate ($MgSO_4 + 7H_2O$)
is used.

1. *Dry Reactions.*—Heated on platinum wire in the zone
of fusion, moistened with cobaltic nitrate, and again heated,
a pink-coloured mass is obtained.

2. *Wet Reactions.*—Dissolve the salt in water, and test
the solution as follows :—

(*a*) Ammonium carbonate gives no precipitate.

(*b* Ammonia precipitates white magnesium hydrate,
$Mg(HO)_2$, soluble in ammonium chloride.

(*c*) Disodium-hydrogen phosphate (Na_2HPO_4), in presence of ammonia and ammonium chloride, precipitates white
crystalline magnesium-ammonium phosphate, $Mg(NH_4)PO_4$.
In very dilute solutions the precipitation takes place only
after some time, and is hastened by stirring the liquid with
a glass rod. The presence of ammonium carbonate does
not interfere with this reaction.

(*d*) Microcosmic salt, $Na(NH_4)HPO_4 + 4H_2O$, produces in hot solutions the same precipitate.

II. POTASSIUM.—Potassium nitrate (KNO_3) is used.

1. *Dry Reactions.*—A crimson-violet colour is imparted to the flame, appearing more blue when viewed through the blue glass

In the spectroscope potassium salts show two lines, one appearing in the outermost red, the other in the violet end of the spectrum. Sodium salts do not interfere with this reaction. (See chart of Spectra.)

2. *Wet Reactions.*—Dissolve the salt in water, and test as follows :—

(*a*) Platinum tetrachloride ($PtCl_4$) produces a yellow crystalline precipitate ($PtCl_4.2KCl$) insoluble in alcohol.

(*b*) Sodium-hydrogen tartrate ($NaHC_4H_4O_6$) precipitates, from strong solutions, white potassium-hydrogen tartrate ($KHC_4H_4O_6$).

These precipitates (*a* and *b*) are more readily formed when the inside of the test-tube is rubbed with a glass rod.

III. SODIUM.—Common salt (NaCl) is used.

1. *Dry Reactions.*—Colours the flame intensely yellow. Viewed through the blue glass, the yellow rays are cut off ; hence if the flame caused by a mixture of sodium and potassium salts be viewed through the blue glass, the blue-violet potassium flame is alone visible.

Viewed through the green glass, the sodium flame appears orange-yellow.

If a piece of paper coated with red mercuric iodide be held close to the yellow sodium flame, the red colour appears changed to white or faint fawn colour.

In the spectroscope sodium salts exhibit a distinct yellow line. (See chart of Spectra.)

2. *Wet Reactions.*—All the sodium salts with scarcely any exception are soluble in water ; there is, therefore, no ordinary reagent which may be used to precipitate the compounds of this metal from their solutions.

K

IV. AMMONIUM.—(Hypothetical radicle, NH_4).

Dissolve a little ammonium chloride in water, and apply the following tests :—

'*a*) Platinum tetrachloride throws down yellow crystalline ammonium-platinic chloride ($2NH_4Cl.PtCl_4$), insoluble in alcohol.

(*b*) Sodium-hydrogen tartrate gives a white crystalline precipitate of ammonium-hydrogen tartrate ($NH_4HC_4H_4O_6$), soluble in a large quantity of water, and in ammonia.

(*c*) Nessler's solution gives a brown precipitate with solutions of ammonium salts ; when these are present in exceedingly minute quantities, a brown colour only is formed.

(*d*) Caustic soda, or potash, heated with ammonium salts decomposes them, ammonia (NH_3) being evolved. (See Lesson VIII., p. 31.)

Separation of Group VI.—Heat a portion of the solution with caustic soda ; ammonia is evolved.

Test in another part of the solution for magnesium by adding ammonia and sodium phosphate (see I. (*c*), p. 128).

Evaporate the rest of the solution to dryness, ignite the residue until all the ammonium salts are volatilised, dissolve the residue in water, again evaporate to dryness, add a few crystals of oxalic acid and ignite. On adding water and filtering, the magnesia remains in the insoluble portion as MgO; the filtrate is evaporated to dryness and a portion tested for sodium by the flame test (see p. 129).

Dissolve the residue in water, and test for potassium by platinium tetrachloride, II. (*a*.), p. 129.

To illustrate the systematic application of the foregoing separations, we will take as example an actually occurring substance, viz. Fahl ore. This mineral may contain silver, mercury, lead, copper, bismuth, arsenic, antimony, iron, zinc, and sulphur, together with sand or gangue. Few specimens contain all these metals. A quantity of the mineral is reduced to fine powder by crushing in a steel mortar, and then pulverising it in an agate mortar. A portion of the sample is then treated in a small beaker glass with strong nitric acid until the residue is white, or nearly so. During this process small particles of sulphur will probably separate and remain undissolved in the acid ; these are yellow in colour, very porous, and generally float towards the surface of the liquid. After filtering from the sulphur and undissolved gangue, which is reserved for further examination, the nitric acid solution is evaporated to a small bulk, diluted with water, and hydrochloric acid added ; the ensuing precipitate is collected (the filtrate being labelled F1. and preserved), washed, boiled in water, filtered, and the filtrate tested for lead by sulphuric acid (III. (*b*), p. 110). The portion insoluble in hot water is dissolved in ammonia, and to the solution nitric acid is added; a white precipitate is indicative of the presence of silver (I. (*a*) p. 109). The filtrate labelled F1. is boiled down until the greater part of the free acid is driven off, diluted with water, and saturated with sulphuretted hydrogen gas.

The precipitated sulphides of Group II. are collected on a filter (filtrate labelled F2.), quickly washed with water containing a little sulphuretted hydrogen (to prevent oxidation of the sulphides), removed to a test-tube, and gently warmed with a little yellow ammonium sulphide and water. Arsenic and antimony sulphides are thus dissolved. The solution is separated from the residue, labelled F3., and preserved. The residue, after being *thoroughly* washed with hot water until free from chlorine,* is boiled in nitric acid, the solution diluted and filtered (the filtrate labelled F4.). The insoluble portion is dissolved in aqua regia, the liquid boiled down, diluted, and tested for mercury by means of stannous chloride (see I. (*b*), p. 111). The filtrate F4. is now

---

* A small quantity of the washings collected separately is tested by adding one drop of nitric acid and one of silver nitrate. If no white precipitate (AgCl) forms, the washing is complete.

boiled down to a small bulk, and a little dilute sulphuric acid (which must be free from lead) added. If a white precipitate forms after some time, it is lead sulphate (see p. 110). Confirm the presence of lead by reduction on the charcoal splinter (see p. 95).

Having filtered from the precipitated lead sulphate, add ammonia in excess ; a blue colour shows the presence of copper. If a white precipitate forms, collect and wash it, dissolve it in one drop of hydrochloric acid, and add water ; a turbidity indicates bismuth (compare II. (*c*), p. 112)

Acidify the blue liquid with hydrochloric acid, and saturate it with sulphuretted hydrogen gas. Collect the resulting precipitate, wash it well and rapidly, and heat it with dilute sulphuric acid (1 part of acid to 5 of water). Filter from undecomposed copper sulphide, and test the filtrate for cadmium by means of sulphuretted hydrogen.

The filtrate labelled F3. is now acidified with hydrochloric acid, the precipitate which ensues is collected, well washed, very gently warmed with a strong solution of ammonium carbonate, and the insoluble portion collected on a filter; the filtrate is acidified with hydrochloric acid, any yellow precipitate which may form is tested on the charcoal splinter for arsenic (p. 99). The portion insoluble in ammonium carbonate, after being thoroughly washed, is dissolved by boiling in strong hydrochloric acid, the solution diluted with water, placed in a platinum basin, and a stick of pure zinc placed in the liquid. After a little time a black deposit appears on the platinum if antimony is present. Pour off the liquid, wash the metallic antimony into a test-tube, allow it to settle, pour away the water, and dissolve the metal in a few drops of nitric acid. Now dilute the solution, and saturate it with sulphuretted hydrogen gas; an orange-coloured precipitate is antimonious sulphide. (Confirm by match reaction, p. 99.)

The filtrate F2. contains the metals of Groups III. and IV. It is boiled until every trace of sulphuretted hydrogen is expelled ; that is, until a piece of bibulous paper moistened with lead acetate solution remains unchanged in colour when held over the boiling liquid. A few crystals of potassium chlorate are added, and the boiling continued until the fluid assumes a deep yellow colour. If any precipitate of sulphur at first forms, it disappears, being oxidised by this treatment with potassium chlorate.

Ammonium chloride and ammonia are now added, and the resulting precipitate collected and washed. A portion is dissolved in hydrochloric acid, and tested for iron by means of potassium ferrocyanide (I. (*c*), p. 119). Another portion is tested on the charcoal splinter for iron (p. 98). The filtrate from the precipitate by ammonia is mixed with a little ammonium sulphide; if a white precipitate forms, it is zinc sulphide. Confirm by the flame reaction (III., p. 124).

The presence of sulphur was indicated when the ore was dissolved in nitric acid (p. 131). To confirm it, use the test with the charcoal splinter described on p. 98.

That portion of the ore which was found to be insoluble in nitric acid (see p. 131) is dried, removed to a porcelain crucible, and fused with about three times its weight of a mixture of sodium and potassium carbonates ; the fused mass is boiled in dilute hydrochloric acid, and the solution evaporated to dryness ; the residue is drenched with hydrochloric acid, hot water added, and the liquid filtered ; the insoluble part is tested for silver by the flame test (see p. 109). To the filtrate ammonia in excess is added. If any precipitate forms it is collected, washed, boiled in caustic soda, the solution filtered, and the filtrate tested for alumina by acidulating with nitric acid and adding ammonium carbonate. The part insoluble in caustic soda is dissolved in hydrochloric acid, and tested for iron (I. (*c*), p. 119). The filtrate separated from the precipitate by ammonia (*vide supra*) is tested for calcium by adding ammonium oxalate (III. (*d*), p. 127). If any precipitate forms it is collected, and the filtrate tested for magnesium by means of sodium phosphate (see I. (*c*), p. 128).

B. *REACTIONS OF THE INORGANIC ACIDS AND ACID RADICLES (INCLUDING OXALIC ACID).*

I. SULPHURIC ACID (H_2SO_4) AND SULPHATES.—Use a solution of sodium sulphate in water.

(*a*) Barium chloride ($BaCl_2$) produces a white precipitate of barium sulphate ($BaSO_4$), insoluble in hydrochloric and nitric acids.*

(*b*) Sulphates are reduced to sulphides when heated on the charcoal splinter (see p. 98).

(*c*) Free sulphuric acid is distinguished from a sulphate by its charring action on cane sugar. A solution containing free sulphuric acid, when evaporated to dryness with a few fragments of sugar, leaves a black charred residue.

To detect very small quantities of free sulphuric acid, a few drops of a solution of cane sugar (1 part of sugar to 30 of water) are placed in a small porcelain basin, one drop of the suspected liquid added, and the basin heated on the water-bath ; a black stain is produced on the basin if the liquid contain so little as 1 part of free acid in 300 parts of water.

Small quantities of sulphur may be detected by heating the dry substance with a little piece of sodium in a hard glass tube, treating the residue with distilled water, and testing the solution—

(1) With sodium nitro-prusside, when a violet colouration ensues.

(2) With ammonium molybdate dissolved in hydrochloric acid, when a blue colour is produced.

The presence of sulphur in a single hair may be detected by the latter test. (Schlossberger.)

---

* The presence of 'glacial phosphoric acid' or of sodium pyrophosphate dissolved in hydrochloric acid interferes with the barium chloride test for sulphuric acid ; the barium sulphate is only thrown down upon heating and then appears as a semi-transparent flocculent precipitate. (Spiller.)

II. Hydrofluosilicic Acid (H_2SiF_6) and Silico-fluorides.—For the preparation of this acid, see Lesson XVI.,
p. 58.

(*a*) Barium chloride precipitates white crystalline barium
silico-fluoride ($BaSiF_6$).

(*b*) Potassium chloride produces a transparent gelatinous
precipitate of potassium silico-fluoride (K_2SiF_6).

(*c*) Silico-fluorides, when warmed with concentrated
sulphuric acid, evolve silicon fluoride, which etches glass
(see p. 59).

III. Phosphoric Acid (H_3PO_4) and Phosphates.—
There are three modifications of this acid.

A. *Orthophosphoric Acid.*—Use a solution of sodium
hydrogen phosphate (Na_2HPO_4) in water.

(*a*) Magnesium sulphate ($MgSO_4$), with previous addition of ammonium chloride and ammonia, throws down a
white crystalline precipitate of magnesium-ammonium phosphate, $Mg(NH_4)PO_4$. Agitation of the liquid promotes the
formation of this precipitate.

(*b*) Barium chloride produces a white precipitate of
barium-hydrogen phosphate ($BaHPO_4$), soluble in nitric and
hydrochloric acids.

(*c*) Silver nitrate throws down yellow silver phosphate
(Ag_3PO_4), soluble in nitric acid and in ammonia. Filter off
the liquid from the precipitated silver phosphate, and test it
with litmus paper ; its reaction is acid.

The liquid filtered from the same precipitate, when
formed in a solution of trisodium phosphate, is neutral to
test-paper.

In the first case nitric acid is produced, as shown
below :—

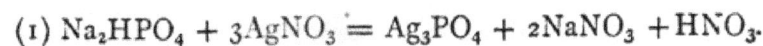

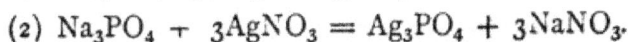

(1) $Na_2HPO_4 + 3AgNO_3 = Ag_3PO_4 + 2NaNO_3 + HNO_3.$

(2) $Na_3PO_4 + 3AgNO_3 = Ag_3PO_4 + 3NaNO_3.$

(*d*) One drop of hydrochloric acid is added, then sodium acetate ($NaC_2H_3O_2$) in excess, and finally *one or two drops* of ferric chloride (Fe_2Cl_6); a yellowish-white gelatinous precipitate of ferric phosphate ($FePO_4$) is formed. This precipitate is soluble in excess of ferric chloride, ferric acetate being formed.

(*e*) Ammonium molybdate, ($NH_4)_2MoO_4$, in nitric acid produces a yellow precipitate, consisting of molybdic acid, with varying amounts of phosphoric acid and ammonia. This precipitate is soluble in ammonia, also in an excess of phosphoric acid or of a phosphate.

The presence of very minute quantities of phosphates may be detected by igniting the substance to be tested, crushing it to fine powder, placing a little of it in a small glass tube, about as thick as a straw, in which a little piece of sodium has been already placed, and gradually heating the tube to redness. Sodium phosphide is formed. After cooling, the fused mass is removed by breaking the tube, placed on a little piece of glass, and moistened with one drop of water, when the very characteristic odour of phosphoretted hydrogen (PH_3) is perceptible.

B. *Pyrophosphoric Acid.*—Heat a little sodium phosphate nearly to redness ; cool, and dissolve in water.

From this solution, which contains sodium pyrophosphate ($Na_4P_2O_7$)—

(*a*) Silver nitrate precipitates white silver pyrophosphate ($Ag_4P_2O_7$), soluble in nitric acid, and also in ammonia.

(*b*) Albumen (white of egg) produces no precipitate.

C. *Metaphosphoric Acid.*—Heat to redness a quantity of ammonium-sodium-hydrogen phosphate, ($NH_4)NaHPO_4$, commonly called microcosmic salt ; dissolve the residue in cold water, and to this solution of sodium metaphosphate ($NaPO_3$) add—

(*a*) Silver nitrate, which precipitates white silver metaphosphate ($AgPO_3$).

(*b*) Albumen, with the addition of acetic acid, is coagulated by this solution, a white flocculent precipitate being produced.

(For detection of phosphorus in organic compounds, see p. 182.)

IV. BORIC ACID (H_3BO_3) AND BORATES.—A solution of borax ($Na_2B_4O_7$) in water is used.

(*a*) Barium chloride precipitates white barium borate $Ba(BO_2)_2$, soluble in hydrochloric acid.

(*b*) Sulphuric acid, when added to a hot concentrated solution of borax, causes a precipitate of boric acid.

(*c*) Acidulate with sulphuric acid, add alcohol, and ignite the mixture ; the flame appears green, especially at the edges. The colour is best seen by repeatedly extinguishing and relighting the flame.

(*d*) Acidify with hydrochloric acid ; immerse a strip of turmeric paper in the solution, withdraw and gently warm it ; the paper acquires a characteristic brown tint. On moistening this paper with a solution of an alkali, the colour changes to bluish-black.

(*e*) Borates, when mixed with 1 part of fluor-spar (CaF_2) and 5 parts of potassium-hydrogen sulphate ($KHSO_4$), and the mixture heated on the platinum loop in the reducing flame of the Bunsen lamp, tinge the flame green. This is owing to the formation of volatile boron fluoride (BF_3) ; the green colour is visible only for a few seconds, and is best seen by viewing the flame against a black background.

By this test small quantities of boron may be detected in minerals. Two experiments should be made—one with the reagents alone, another after the addition of the substance to be tested.

V. OXALIC ACID ($H_2C_2O_4$) AND OXALATES.—A solution of ammonium oxalate, $(NH_4)_2C_2O_4$, in water is used.

(*a*) Barium chloride precipitates white barium oxalate (BaC_2O_4), soluble in hydrochloric and nitric acids.

(*b*) Silver nitrate produces a white precipitate of silver oxalate ($Ag_2C_2O_4$), soluble in nitric acid and in ammonia.

(*c*) Calcium chloride produces a white precipitate of calcium oxalate, insoluble in acetic acid.

This substance by gentle ignition is converted into calcium carbonate, which effervesces on the addition of an acid.

(*d*) On heating oxalates with strong sulphuric acid, carbon monoxide and dioxide are evolved, and the residue is *not* blackened ; the former gas burns with a pale lavender flame on bringing a light to the mouth of the tube ; the presence of the latter is recognised by decanting the gaseous fumes into a test-tube containing clear lime-water. (See Lesson IX., p. 37.)

VI. HYDROFLUORIC ACID (HF) AND FLUORIDES. — (Fluorspar is used for the insoluble fluorides.)

(*a*) Barium chloride precipitates white barium fluoride (BaF_2), soluble in hydrochloric acid.

(*b*) Fluorides, when finely powdered, moistened with strong sulphuric acid, and *very gently* warmed in a leaden or platinum dish, evolve hydrofluoric acid. The capsule is covered with a watch-glass, having a thin film of wax spread over its convex side, letters being traced through the wax with a pin or other pointed instrument. Wherever the glass is exposed to the action of the hydrofluoric acid, it is etched ; this etching is made apparent by removing the wax (by gentle warming) and breathing on the glass. (Compare Lesson XVI., p. 59.)

(*c*) Fluorides containing silica evolve silicon fluoride (SiF_4) when heated with sulphuric acid. If this gas be conducted through a wet glass tube, it is decomposed, hydrated silicic acid being deposited, which, on drying the tube, appears as a white powder on the glass.

Small quantities of fluorine may be detected (1) by mixing the substance with a few drops of strong sulphuric acid in a watch-glass, heating until dry, and washing off the

residue with water. A dull spot appears on the glass. Or
(2) the substance (mixed with silica, if that body is not
already present) is heated with strong sulphuric acid in a
large test-tube, the evolved gas led into ammoniacal water
in another tube, the liquid evaporated to dryness, the residue
heated with water and filtered, the filtrate again evaporated
to dryness, and the residue tested for fluorine as directed in
(*b*) above. (Wilson.)

VII. Sulphurous Acid (H_2SO_3) and Sulphites.—
A solution of sodium sulphite in water is used.

(*a*) Barium chloride precipitates white barium sulphite
($BaSO_3$), soluble in hydrochloric acid with evolution of
sulphur dioxide.

(*b*) Silver nitrate throws down white silver sulphite
(Ag_2SO_3). On heating, this precipitate becomes dark grey
or black, metallic silver being produced :

$$Ag_2SO_3 + H_2O = H_2SO_4 + Ag_2.$$

(*c*) Treated with zinc and hydrochloric acid, sulphuretted
hydrogen is produced : very small quantities of the acid
may be thus detected.

(*d*) Free sulphurous acid (or sulphur dioxide in a gaseous
mixture) may be detected by its power of liberating iodine
from iodic acid. Paper moistened with starch paste containing a small quantity of this acid is rendered blue by the
action of sulphurous acid.

(*e*) Acidulate with hydrochloric acid and pass sulphuretted hydrogen through the liquid ; sulphur is deposited,
pentathionic acid ($H_2S_5O_6$) being simultaneously produced.

VIII. Silicic Acid—$Si(HO)_4$—and Silicates.

(*a*) The alkaline silicates are soluble in water ; their
solutions are decomposed by acids, or by heating with
ammonium chloride or carbonate, silicic acid separating as
a gelatinous precipitate. If the solution is very dilute it

must be evaporated to dryness, when insoluble silicon dioxide (SiO₂) remains. This substance—

(*b*) When heated in the platinum loop with a *small quantity* of sodium carbonate yields a clear colourless bead, carbon dioxide escaping with effervescence. When heated with microcosmic salt the silica remains undissolved, forming the so-called silica skeleton bead.

For the purposes of analyses silicates may be divided into two classes—

(I.) Silicates decomposable by acids.

(II.) Silicates undecomposable by acids.

Silicates of the first class are analysed by evaporating the specimen, *reduced to very fine powder*, with hydrochloric acid, to complete dryness, adding dilute acid, and boiling with addition of water. On filtering the insoluble silica remains on the filter (test in the bead before the Bunsen flame), while the bases may be detected in the filtrate.

Silicates of Class II. are reduced to *fine* powder, and intimately mixed in a platinum crucible with three or four parts of pure sodium carbonate. The mixture is fused over the blowpipe until a liquid mass is obtained, which, after cooling, is boiled with hydrochloric acid, and subsequently evaporated to complete dryness; the remainder of the process is identical with that described above.

In order to detect the presence of alkalies in a silicate, the substance is fused with about three-fourths of its own weight of powdered ammonium chloride and three or four times its weight of pure calcium carbonate, (prepared by dissolving marble in hydrochloric acid, adding milk of lime to alkaline reaction, filtering, and precipitating the filtrate, after heating, with ammonium carbonate). After fusion the mass is boiled with water for a considerable time, filtered, the lime precipitated by means of ammonium carbonate, and the filtrate from this precipitate evaporated to dryness, and the residue ignited. This residue is then tested for sodium, potassium, and lithium. Fluorine may be detected by the method (2) on p. 139. Titanium by the method on p. 175.

IX. CARBONIC ACID (H_2CO_3) AND CARBONATES.—Sodium carbonate may be used.

(*a*) Barium chloride precipitates, from neutral solutions, barium carbonate ($BaCO_3$), soluble in acids with effervescence.

(*b*) Carbonates when pulverised and treated with hydrochloric acid evolve carbon dioxide (CO_2). If the fumes resulting from the decomposition of a carbonate be decanted into a test-tube containing clear lime water, a white precipitate of calcium carbonate is produced. These fumes do not form a white cloud with hydrochloric acid.

X. IODIC ACID (HIO_3) AND IODATES.—Use a solution of the acid in water.

(*a*) Barium chloride precipitates white barium iodate, $Ba(IO_3)_2$, soluble in nitric acid.

(*b*) Sulphur dioxide precipitates iodine ; but when excess of the reagent is added, hydriodic acid is produced.

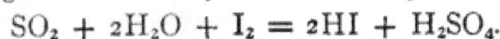

$$SO_2 + 2H_2O + I_2 = 2HI + H_2SO_4.$$

To detect iodates in the presence of iodides, acidulate the solution with sulphuric acid and add a little starch paste; if a blue colour is produced, an iodate is present (see VII. (*d*), p. 139).

$$HIO_3 + 5HI = 3H_2O + 3I_2.$$

XI. THIOSULPHURIC ACID ($H_2S_2O_3$) AND THIOSULPHATES.—Sodium thiosulphate is used.

(*a*) Hydrochloric acid, on gently warming, produces a precipitate of sulphur, accompanied with the evolution of sulphur dioxide.

(*b*) Lead acetate produces a white precipitate of lead thiosulphate (PbS_2O_3). On boiling, this precipitate becomes black, being changed to a compound having the composition of $PbS.PbS_2O_3 + H_2O$.

(*c*) Ferric chloride produces a reddish-violet colouration, which disappears on heating or on standing for some time.

(*d*) Silver nitrate precipitates white silver thiosulphate ($Ag_2S_2O_3$), soluble in nitric acid. This precipitate rapidly blackens in colour, owing to the formation of silver sulphide (Ag_2S).

XII. HYDROCHLORIC ACID (HCl) AND CHLORIDES.— A solution of sodium chloride in water is used.

(*a*) Silver nitrate produces a white curdy precipitate of silver chloride (AgCl), which darkens on exposure to the light. This precipitate is insoluble in nitric acid, but dissolves easily in ammonia.

(*b*) Chlorides evolve chlorine when heated with sulphuric acid and manganese dioxide.

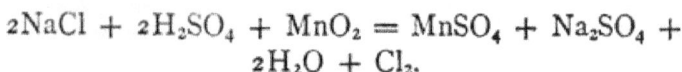

$$2NaCl + 2H_2SO_4 + MnO_2 = MnSO_4 + Na_2SO_4 + 2H_2O + Cl_2.$$

(*c*) Dry chlorides when heated in a retort with sulphuric acid and potassium bichromate yield a dark red distillate of chromyl dichloride (CrO_2Cl_2), which on addition of ammonia gives a yellow solution containing ammonium chromate :

$$CrO_2Cl_2 + 2NH_4HO = (NH_4)_2CrO_4 + 2HCl.$$

If a soluble lead salt be now added, a yellow precipitate of lead chromate ($PbCrO_4$) ensues ; or a yellowish colour is produced on adding an acid.

(*d*) If a little cupric oxide be dissolved in a bead of microcosmic salt in the oxidising flame, a small quantity of a chloride added, and the bead brought into the reducing flame, a bluish-coloured flame, inclining to purple, is seen around the bead. Viewed through the blue glass the flame appears azure blue inclining to green.

XIII. HYDROBROMIC ACID (HBr) AND BROMIDES.—A solution of potassium bromide in water is used.

(*a*) Silver nitrate precipitates white silver bromide (AgBr), insoluble in dilute nitric acid; soluble in ammonia, potassium cyanide, or sodium thiosulphate.

(*b*) Bromides, when heated with sulphuric acid and manganese dioxide, evolve bromine, which is recognised by its red colour and by its power of turning starch paste yellow or yellowish red.

This test is best applied by gently heating the mixture in a small beaker, the mouth of which is covered with a watch-glass, to the lower side of which is attached a slip of paper moistened with starch paste and strewed over with dry starch. (Fresenius.)

(*c*) Dry bromides, when heated in a retort with sulphuric acid and potassium bichromate, yield a dark red distillate, which consists of pure bromine. The addition of excess of ammonia to this liquid decolourises it completely.

This test serves to discriminate between bromides and chlorides, and also to detect either of these salts in presence of the other.

(*d*) Chlorine in the gaseous state or in solution decomposes solutions of bromides, bromine being liberated, and colouring the liquid more or less yellow. On shaking the tube after adding a few drops of carbon disulphide, the bromine is concentrated in this liquid, to which it imparts a reddish-brown colour. Excess of chlorine must be avoided, else the colour is discharged.

(*e*) Bromides treated as described for chlorides on p. 142 (*d*), give a blue flame, appearing greenish at the edges.

XIV. HYDRIODIC ACID (HI) AND IODIDES.—Potassium iodide is used.

(*a*) Silver nitrate precipitates pale yellow silver iodide (AgI), insoluble in dilute nitric acid, almost insoluble in ammonia, but easily dissolved by potassium cyanide or sodium thiosulphate.

(*b*) Bromine water liberates iodine from iodides ; on shaking with carbon disulphide the iodine is concentrated in this liquid, forming a violet-coloured solution.

This appears to be the most delicate test for iodine. The bromine water should be rather dilute, a few drops are added, then a couple of drops of carbon disulphide, followed by further *cautious* additions of bromine water, and *gentle* agitation of the liquid until the violet colouration is visible.

(*c*) A solution of nitrous fumes in sulphuric acid likewise liberates iodine from iodides ; chlorine water causes the same reaction. On the addition of a little starch paste a deep blue colour is produced, which disappears on heating the liquid. If an excess of chlorine water is used, no blue colour is produced on adding starch paste (owing to the formation of colourless iodine chloride); but if a reducing agent, as stannous chloride, be now added, the blue iodide of starch is re-formed.

(*d*) When solutions of the iodides are mixed with a little starch paste, acidulated with hydrochloric acid, and subjected to electrolysis, the characteristic blue colour is produced, especially around the negative pole. (Pelloggio.)

(*e*) Boiled with ferric chloride, iodine is evolved :

$$Fe_2Cl_6 + 2KI = 2KCl + 2FeCl_2 + I_2.$$

(*f*) Dry iodides when heated in a glass tube closed at one end, with a little strong sulphuric acid, evolve gaseous iodine, which fills the upper part of the tube, and is recognised by its violet colour.

(*g*) With the microcosmic bead, as directed for chlorides (p. 142 (*d*)), iodides give a green colouration to the flame.

Very small quantities of iodine may be detected by adding a little starch paste to the solution to be tested, a few drops of a dilute solution of potassium bichromate, just sufficient to communicate a pale yellow colour, and finally a drop or two of hydrochloric acid, when a blue colour, or, in extremely dilute solutions, a tawny colour, is apparent.

Chlorine, bromine, and iodine, in organic compounds are detected by heating the finely powdered substance with pure lime in a hard glass tube, dissolving the contents of the tube in dilute nitric acid, and applying the several tests to this solution. The substance, if a liquid, is placed in a little glass bulb, which is set inside the glass tube.

Iodides, bromides, and chlorides, when present together in a solution, may be detected as follows :—

The solution is divided into two portions. To the first solution a drop of a solution of nitrous fumes in sulphuric acid is added, and then a few drops of carbon disulphide, with gentle agitation ; the carbon disulphide assumes a violet colour more or less deep according to the quantity of iodine present. The liquid is decanted, a drop or two of carbon disulphide again added, and then diulte chlorine water, drop by drop, with gentle agitation. The characteristic reddish-yellow colour of bromine soon appears. The second portion of the solution is evaporated to dryness with addition of sodium carbonate, the dry mass fused with potassium bichromate in a porcelain crucible ; if iodine is present, it is driven off. The fused mass is broken into little pieces, which are placed in a retort and heated with concentrated sulphuric acid. Part of the distillate is treated with ammonia ; if it assumes a yellow colour, chromyl dichloride is present. As a confirmatory test, the other part of the distillate is boiled with the addition of alcohol, and a little hydrochloric acid and ammonia added (see tests for chromium, p. 121). If a pale green precipitate forms, the presence of chromium is conclusively proved. Chlorine was, therefore, present in the original mixture. (See XII. (*c*), p. 140.)

XV. HYDROCYANIC ACID (HCN) AND CYANIDES.— A solution of potassium cyanide in water is used.

(*a*) Silver nitrate precipitates white silver cyanide (AgCN), insoluble in nitric acid, soluble in ammonia, also in excess of potassium cyanide and in sodium thiosulphate. Silver cyanide is decomposed on ignition, metallic silver being produced. In this respect it differs from silver chloride.

(*b*) A solution containing a mixture of ferrous and ferric salts (prepared by exposing ferrous sulphate solution to the air for a short time) throws down, from solutions of the cyanides made alkaline by caustic soda, a greenish-blue precipitate consisting of Prussian blue with admixed hydrated iron oxides. On the addition of hydrochloric acid the latter are dissolved, Prussian blue remaining.

(*c*) A few drops of hydrochloric acid are added to a little of the potassium cyanide solution in a small basin, which is covered with another basin to the convex side of which a drop of yellow ammonium sulphide adheres. The lower basin is very gently heated, the upper basin removed, after a minute or two, and excess of ammonium sulphide driven off by *very slightly* warming the basin, and blowing on the drop of liquid. Ammonium sulphocyanide, $(NH_4)CyS$, is thus formed. When the basin is quite cool, add one drop of ferric chloride to the spot on the dish; a blood-red colouration is produced. (Compare p. 120.)

Analysis of Cyanides.—All cyanides evolve hydrocyanic acid when heated with concentrated hydrochloric acid. The alkaline cyanides may be dissolved in water, and the foregoing tests applied. Mercuric cyanide $(HgCy_2)$ is dissolved in water, the mercury removed as sulphide by precipitation with sulphuretted hydrogen, and the filtrate tested as above. The other simple cyanides may be fused in a porcelain crucible with sodium carbonate, the fused mass dissolved in water, and to this solution the usual tests applied.

The insoluble double cyanides (Prussian blue, Turnbull's blue, &c.) must be boiled with caustic soda and filtered. The precipitate contains the bases (ferric oxide, &c.), which are separated in the usual manner. Potassium ferro- or ferri-cyanide goes into solution along with the bases soluble in caustic soda which may be present (lead, zinc, or aluminium oxides). Sulphuretted hydrogen is passed through the alkaline liquid to remove these metals, and the filtrate, divided into two parts, is tested for potassium ferrocyanide with ferric chloride (see (*h*), p. 120), and for ferri-cyanide with ferrous sulphate (see (*d*), p. 119).

The alkalies which are present in certain insoluble cyanides may be detected by heating with three volumes of concentrated sulphuric acid and one volume of water, until excess of acid is driven off. The residue is dissolved in hydrochloric acid, the heavy metals which may be present removed, and the alkalies detected by the usual methods.

XVI. NITROUS ACID—HNO_2—AND NITRITES.—Potassium nitrite, dissolved in water, is used.

(*a*) Silver nitrate precipitates white silver nitrite ($AgNO_2$), soluble in a large excess of water.

(*b*) Ferrous sulphate produces a brownish-black colouration, owing to the formation of nitric oxide, and solution of the same in the ferrous sulphate (compare reaction (*a*) under No. XIX.).

(*c*) Mercurous salts are reduced by nitrites, grey metallic mercury being deposited.

(*d*) Add a little potassium iodide, a drop of starch paste, and a few drops of dilute sulphuric acid ; a blue colour ensues, owing to the decomposition of the potassium iodide, and the action of the liberated iodine on the starch.*

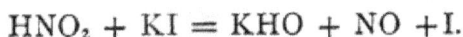

$$HNO_2 + KI = KHO + NO + I.$$

The reactions (*b*), (*c*), and (*a*) serve to distinguish between nitrites and nitrates.

XVII. HYPOCHLOROUS ACID—HClO—AND HYPOCHLORITES.—A solution of sodium hypochlorite in water is used.

(*a*) Silver nitrate produces a white precipitate of silver chloride.

(*b*) Lead nitrate throws down a precipitate which, at first

---

* The potassium iodide used for this test must be free from iodate. If on addition of sulphuric acid and starch paste to the iodide, a blue colour is produced, it contains iodate (see reaction (c) for iodic acid p. 141).

white, changes to red, and then to brown, from formation of lead dioxide (PbO_2).

(*c*) Manganese chloride ($MnCl_2$) precipitates dark brown hydrated manganese dioxide ($MnO(HO)_2$).

(*d*) If a hypochlorite be treated with water, a few crystals of cobaltic nitrate added, and the mixture gently heated, oxygen is evolved.

(*e*) Hypochlorites are decomposed by dilute hydrochloric acid with evolution of chlorine.

XVIII. Hydrosulphuric Acid — H_2S — and Sulphides.—A solution of sodium sulphide is used.

(*a*) Silver nitrate precipitates black silver sulphide (Ag_2S).

(*b*) Lead acetate throws down black lead sulphide (PbS).

(*c*) Sodium nitro-prusside ($Na_2Fe(NO)Cy_5$) produces a deep but transient violet-red colour. This test is only applicable to sulphides of the alkali metals. Before applying it to other sulphides caustic soda or potash must be added to the solution to be tested.

(*d*) Most sulphides are decomposed by hydrochloric or sulphuric acid, hydrosulphuric acid being liberated. This gas is recognised by its odour, and by its blackening action on paper moistened with a solution of a lead salt.

(*e*) When sulphides are fused on the platinum loop with a little solid caustic soda or potash, sodium or potassium sulphide is formed. These sulphides are soluble in water. If a silver coin be immersed in this solution it is blackened, owing to the formation of silver sulphide.

Sulphates are reduced to sulphides only by fusion in presence of reducing agents (see p. 98).

To detect sulphides in the presence of sulphites and thiosulphates proceed as follows :—

A portion of the liquid is tested for hydrosulphuric acid by tests (*c*) and (*d*) above. To another part of the solution zinc sulphate solution, made strongly alkaline by means of caustic potash, is added, the precipitate of zinc sulphide removed by filtration, and the filtrate divided into two portions. In one of these thiosulphuric acid is recognised by the evolution of sulphur dioxide, accompanied by the deposition of sulphur, on very gently warming with hydrochloric acid. In the other portion, very slightly acidified by acetic acid, sulphurous acid is recognised by the reddish colour produced on the addition of a tolerably large excess of zinc sulphate and a few drops of sodium nitro-prusside. This reaction is made more delicate by adding a very little potassium ferrocyanide.

XIX. NITRIC ACID—HNO_3—AND NITRATES.—Potassium nitrate (KNO_3), dissolved in water, is used.

(*a*) Add a solution of ferrous sulphate, mix the liquids, and carefully pour a few drops of strong sulphuric acid down the side of the test tube ; a brown ring is formed at the point of contact of the two fluids.

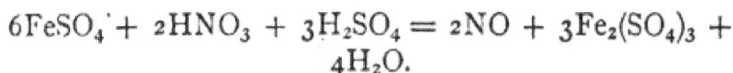

$$6FeSO_4 + 2HNO_3 + 3H_2SO_4 = 2NO + 3Fe_2(SO_4)_3 + 4H_2O.$$

(Compare lesson V. p. 22.)

In applying this test to nitrates which are decomposed by sulphuric acid, giving rise to insoluble sulphates, *e. g.*, barium nitrate, the solution is mixed with an equal volume of sulphuric acid, the precipitate allowed to settle, and a solution of ferrous sulphate then carefully added.

(*b*) Add a little indigo solution, a few drops of hydrochloric acid, and boil ; the blue colour of the indigo disappears, *isatin* being formed.

$$3C_8H_5NO + 2HNO_3 = 3C_8H_5NO_2 + 2NO + H_2O.$$

(*c*) Add a little dilute sulphuric acid, a few drops of potassium iodide solution, and a little starch paste. Immerse a strip of zinc foil in the liquid, the nitric is reduced to nitrous acid, which liberates iodine from potassium iodide (compare reaction (*d*) for nitrites, p. 147).

(*d*) Small quantities of nitric acid may be detected by

evaporating the solution to dryness, moistening with a drop of *pure* sulphuric acid, and moving about a fragment of brucine in the liquid, when a rose-red colouration ensues. By using 1 c.c. of water this test will detect nitric acid in a water in which the quantity does not exceed ·1 Mgm. per litre (Nicholson).

To detect nitrates in the presence of chlorates the solution is poured into a little test tube containing a few pieces of

FIG. 51.

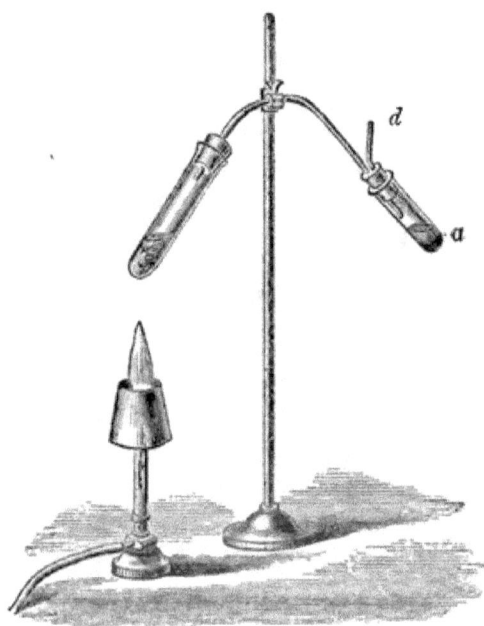

zinc foil, coated with spongy copper.* This tube is connected by means of a little bit of glass tubing with another smaller tube (see fig. 51). The second tube, *d*, serves to establish connection with the outer air.

---

* A few pieces of zinc foil, cleaned by immersion in sulphuric acid and washing with distilled water, and bent so as to present a large surface in a small bulk, are placed in the test tube, and covered with a moderately warm solution of cupric sulphate : after the expiry of ten minutes or so the liquid is poured off, and the foil, which is now coated with spongy copper, carefully washed with distilled water.

A is heated until the greater part of the liquid has distilled over into *a*. In this reaction nascent hydrogen is produced, and by its action the nitric acid is reduced to ammonia ($HNO_3 + H_8 = NH_3 + 3H_2O$). In the distillate ammonia is detected by its reaction with Nessler's solution (see test (*c*) for ammonia, p. 130).

The chlorate present is also reduced, a chloride being formed, which remains in solution in the tube A. On pouring off the liquid from this tube and testing it with silver nitrate, the reactions for chlorine are obtained (see p. 142). In the original solution the chloric acid may be detected by test (*a*) (*vide infra*).

Nitrogen in organic bodies may be detected by heating the substance in a clean dry test tube, with a small piece of metallic potassium, whereby potassium cyanide is produced. After cooling, water is carefully added, the solution filtered, and cyanogen detected by test (*b*) (p. 146).

XX. CHLORIC ACID—$HClO_3$—AND CHLORATES.—A solution of potassium chlorate in water is used.

(*a*) A few drops of indigo solution are added, the liquid acidulated with sulphuric acid and sulphurous acid (or sodium sulphite dissolved in water), added drop by drop. The blue colour is discharged.

(*b*) Dry chlorates treated with sulphuric acid evolve a yellow explosive gas (Cl_2O_4). In applying this test only a small quantity of the salt must be used, and the mixture kept cold, otherwise a violent explosion may occur.

(*c*) Chlorates when heated evolve oxygen, a chloride being simultaneously formed.

(For detection of chlorates in the presence of nitrates, see p. 150.)

XXI. PERCHLORIC ACID — $HClO_4$ — AND PERCHLORATES.—Potassium chlorate is fused until the evolution of gas nearly ceases, and the mass gets semi-solid. The residue contains potassium perchlorate, which is separated from admixed chlorate and chloride by dissolving out the latter salts in successive small quantities of warm water, the perchlorate being much more insoluble than either of the other salts.

(*a*) Heated with strong sulphuric acid, white fumes of hydrated perchloric acid are evolved.

(*b*) The solution of a perchlorate, mixed with hydrochloric acid, does not discharge the colour of indigo solution.

(*c*) Dry perchlorates when heated evolve oxygen.

C. *REACTIONS OF THE ORGANIC ACIDS.*

The organic acids are in general more difficult to detect than the inorganic acids. The tests here given serve, in most cases, for the identification of the more commonly occurring members of this group of acids.

The only organic acids which we shall notice are—

Tartaric, citric, benzoic, succinic, acetic, and formic acids.

I. TARTARIC ACID ($H_2C_4H_4O_6$).—A solution of sodium or potassium tartrate in water is used.

(*a*) Solutions of the tartrates (if neutral, acidify with acetic acid) give a white precipitate of potassium hydrogen tartrate, $KHC_4H_4O_6$, on the addition of caustic potash.

(*b*) Calcium chloride precipitates white calcium tartrate ($CaC_4H_4O_6$), soluble in acids, also in salts of ammonium. This precipitate is soluble in *cold* caustic potash, but is re-precipitated on boiling.

(*c*) To a solution of a neutral tartrate add silver nitrate; filter off the white precipitate of silver tartrate ($Ag_2C_4H_4O_6$) which forms; dissolve this in ammonia, and warm the solution for some time; a mirror of silver is formed on the test-tube.

II. CITRIC ACID ($H_3C_6H_5O_7$).—A solution of sodium citrate in water is to be used.

(*a*) Calcium chloride produces, on boiling, a white precipitate of calcium citrate, $Ca_3 (C_6H_5O_7)_2$. This precipitate is *insoluble* in caustic potash, but soluble in hydrochloric acid.

(*b*) Lime-water produces a similar precipitate on boiling the liquid.

Tartaric acid in the presence of citric acid may be detected by adding hydrated ferric oxide to the solution, heating the mixture to about 90°, filtering from undissolved ferric oxide, and evaporating the filtrate to a syrupy consistence. If tartaric acid is present, an insoluble basic tartrate is formed and precipitated, ferric citrate remaining in solution.

III. BENZOIC ACID $(HC_7H_5O_2)$.—The ammonium salt, dissolved in water, is used.

(*a*) Calcium chloride gives *no precipitate.*

(*b*) Ferric chloride precipitates (from neutral solutions) ferric benzoate, which is decomposed by ammonia, ferric hydrate remaining as an insoluble residue. This precipitate is soluble in a small quantity of hydrochloric acid, benzoic acid being liberated.

IV. SUCCINIC ACID $(H_2C_4H_4O_4)$.—Ammonium succinate dissolved in water is used.

(*a*) Barium or calcium chloride alone gives no precipitate, but on the addition of alcohol, white barium or calcium succinate $(Ba (or Ca) C_4H_4O_4)$ is precipitated.

(*b*) Ferric chloride gives a brownish-red precipitate of ferric succinate, which behaves like ferric benzoate with ammonia.

V. ACETIC ACID (HC_2H_3O).—A solution of sodium acetate in water is used.

(*a*) Ferric chloride in neutral solutions produces a deep blood-red colour. This solution, on addition of hydrochloric acid, turns yellow, and may be thus distinguished from ferric sulphocyanide. On boiling the red solution, all the iron is precipitated as a basic acetate, the liquid becoming colourless.

(*b*) Dry acetates, heated with arsenious oxide, yield the very foully-smelling cacodyl $\left[As(CH_3)_2\right]$

(*c*) With silver nitrate a white precipitate of silver acetate $(AgC_2H_3O_2)$ is formed.

(*d*) Treated with strong sulphuric acid and alcohol, the characteristic fruity smell of acetic ether $(C_2H_5.C_2H_3O_2)$ is perceived.

VI. FORMIC ACID $(HCHO_2)$.—Sodium formate dissolved in water is used.

(*a*) With ferric chloride the formates behave like the acetates.

(*b*) Heated with silver nitrate, a precipitate of metallic silver is produced. Silver formate is first produced ; this is then reduced to metallic silver $(2AgCHO_2 = 2Ag + CO_2 + HCHO_2)$.

In the event of the solution to be examined containing metals, it should be boiled with excess of sodium carbonate, the filtrate slightly acidified with hydrochloric acid (to remove carbon dioxide), gently heated, again neutralised with ammonia, and calcium chloride added.

D. SYNOPSIS OF ANALYTICAL METHODS.

TABLE A.

HEAT THE SOLID SUBSTANCE IN A GLASS TUBE CLOSED AT ONE END.

It Fuses.	It does not Fuse, but changes Colour.	It evolves Gas.	It forms a Sublimate.	Drops of Water are formed.
Alkaline salts It fuses, and changes colour. Colour white Hot · Cold Bismuth oxide } *yellow . dark yellow* Lead oxide } *yellow . dark red* If the substance blackens, organic matter is present.	Before heating. · While hot. Zinc oxide. *white* . *yellow* Stannic oxide. } *straw. dark yellow* Ferric oxide. } *red . black* Mercuric oxide. } *red . brown red*	1. Gases are colourless and odourless a. *Oxygen.* Easily decomposed peroxides, chlorates and nitrates. b. *Carbon monoxide.* Oxalates. 2. Gases are colourless, but possessed of an odour. a. *Ammonia.* Ammonium salts. b. *Sulphur dioxide.* Sulphites and certain sulphates. c. *Cyanogen.* Cyanogen compounds. 3. Gases are coloured. a. *Brown red—nitrous oxide.* Nitrites and certain nitrates of the heavy metals.	1. *Reddish brown drops.* Sulphur. 2. *White sublimate.* Ammonium salts, mercurous and mercuric chlorides, arsenious and antimonious oxides. 3. *Yellow sublimate.* Mercuric iodide, arsenious sulphide. 4. *Orange yellow sublimate, changing to black on heating.* Mercuric sulphide. 5. *Metallic mirror.* Arsenic. 6. *Metallic globules.* Mercury.	1. *Water of crystallisation, reaction generally neutral.* Easily fusible salts. 2. *Water of combination.* Generally non-fusible salts. a. *Water reacts acid.* Sulphites, chlorides, &c. b. *Water reacts alkaline.* Ammonium salts.

This Table is to be used only as a guide to further Examinations. The reactions herein given may be masked by the presence of a number of metals; therefore, although we may not obtain a distinct reaction for a certain metal, we are not positively to decide that this metal is absent.

TABLE B.

FLAME REACTIONS WITH THE BUNSEN LAMP.

(Refer to page 90.)

A Heated in the cold portion of the flame. The flame is coloured.	B Heated on borax bead in the lower oxidising flame. The bead is coloured.	C Heated on match splinter, in upper reducing flame, &c. Reduced metal is obtained.	D Heated on asbestos in upper reducing flame, films are obtained on porcelain.
1. *Yellow* . . . Sodium	1. *Yellow.* Silver (often opal colour).	1. *Magnetic.* Iron. Nickel. Cobalt.	1. *Metallic - looking spots.* Probably Arsenic. Sulphide film yellow.
2. *Red* . . . . . { Strontium and Lithium	Bismuth (colourless when cold). Lead. Cadmium (only when present in large quantity, colourless when cold).	2. *Malleable.* (Press the globules with the small pestle.) Lead, Tin, Silver, Copper (red colour).	2. *Sooty - like spots.* Antimony. S. F. orange.
3. *Green* . . . { Barium (Thallium)	Uranium(yellow green)	3. *Brittle.* Antimony.	3. *Metallic mirror spreading over the disc.* Mercury. S. F. black.
4. *Bluish green.* Copper	2. *Green.* { Chromium (emerald green). Copper (blue when cold).	(Further tests may be applied by dissolving the reduced metal and applying *wet reaction*, as described on p. 96.	4. *Black film.*
5. *Lavender* . . Potassium	3. *Blue.* Cobalt.	4. Sulphates are thus reduced to sulphides. The reduced mass is to be treated with water, and a silver coin dipped into the solution, a *black stain* of Silver sulphide (Ag₂S) is noticed on the coin.	a. Lead . . S. F. brownish red to black.
6. *Yellowish red* Calcium	4. *Red to Brown.* { Iron (getting gradually colourless on cooling). Nickel (colourless when cold).		b. Cadmium S. F. lemon.
	5. *Amethyst.* Manganese.		c. Zinc . . „ white.
	6. If silicic acid be present, it remains undissolved in the bead.		d. Tin . . „ „

The use of this Table will afford valuable indication as to what metals we are likely to find, but further tests are necessary before deciding if a given metal is or is not present

TABLE C.

SOLUTION OF THE SUBSTANCE.

A. Metals and Alloys.	B. Substances which are neither Metals nor Alloys, and contain no Organic Matter.	C. Substances which are neither Metals nor Alloys, but contain Organic Matter.
Boil with strong nitric acid. (Examine solution after Table D.) Residue (if any) may be I. METALLIC: *Gold* or *Platinum.* Boil in aqua regia and evaporate with hydrochloric acid. Divide the solution into two portions and add to Portion *a.* Ferrous sulphate, precipitate *Gold.* Portion *b.* Potassium chloride, precipitate *Platinum.* II. WHITE and NON-METALLIC: *Tin* or *Antimony* (as oxides). *a.* Everything dissolves. *Antimony* (orange precipitate in solution with sulphuretted hydrogen). *b.* An insoluble substance remains. *Tin.* Confirm by match-reaction in Bunsen flame.	I. Boil a small quantity in fine powder with water. *a. All is dissolved.* Treat more of the substance with water, and examine solution after Table D. *b. An insoluble portion remains.* Filter; evaporate a few drops of filtrate on platinum foil. If there is a considerable residue, treat all the substance with water and proceed with filtrate by Table D; if only slight residue, proceed to next test. II. Part of that which was insoluble in water—or, if the substance was entirely insoluble in water, part of the original substance is boiled with concentrated hydrochloric acid. (Notice if any gases are evolved.) *a. All is dissolved.* Treat more of original substance with hydrochloric acid; and, if Bunsen test has shown no silicic acid, examine solution by Table D. If silicic acid be present, evaporate to dryness; warm residue with concentrated hydrochloric acid, filter from silica, and proceed by Table D. *b. Insoluble matter remains.* Dilute with water and boil again; if anything insoluble still remains, proceed to III. III. Another part of that which was insoluble in water —or, if it is completely insoluble therein, a portion of the original substance— is boiled with concentrated nitric acid. *a. All is dissolved.* Treat more of original substance in same way; evaporate to dryness, if necessary, to get rid of silicic acid, and proceed with solution by Table D. *b. Insoluble remains.* Boil this in aqua regia; if it dissolves, proceed with solution as in III. *a.*; if insoluble, pass to Table O.	The organic matter must be removed, either by treatment with water, by ignition, or by gently warming with concentrated hydrochloric acid, with addition from time to time of potassium chlorate. The particular method to be employed will depend upon the nature of the material under examination. Compound cyanides are most easily decomposed by heating with concentrated sulphuric acid, and then boiling until part of the acid is driven off.

TABLE D.
TREATMENT OF THE SOLUTION.

A	B	C
Treatment of the aqueous solution.	*Treatment of the hydrochloric acid solution.*	*Treatment of the nitric acid solution.*
Proceed with the systematic addition of the group reagents.	Evaporate off the excess of acid, dilute the solution with water, and proceed with the addition of the group reagents, beginning with sulphuretted hydrogen. Silver, lead, and monad mercury, which form chlorides insoluble in water, cannot, of course, be present in the solution.	Evaporate the solution nearly to dryness, and add water. If you have both a hydrochloric and a nitric acid solution, mix them together, and evaporate almost to dryness; add water, and proceed with the addition of the group reagents, commencing with sulphuretted hydrogen. If silver, lead, or mercury be present in the nitric acid solution, they will remain undissolved on treatment with water. The insoluble portion must be treated as directed in Table E.

TABLE E.

EXAMINATION OF PRECIPITATE BY HYDROCHLORIC ACID.

This Precipitate may be obtained on addition of Hydrochloric Acid to a watery solution, or sometimes to a Nitric Acid solution.

The precipitate, after being washed with cold water, is transferred to a test tube, boiled with water and filtered.

Solution. Add sulphuric acid white precipitate = $PbSO_4$. LEAD.

Residue. Treat with ammonia and filter.

   Solution. Add nitric acid white precipitate = AgCl. SILVER.

   Insoluble portion. Blackened = NH_2Hg_2Cl (as mercurous salt). MERCURY.

If the original substance be a liquid, its action upon litmus-paper must be determined. If *acid,* the addition of *water* may precipitate basic bismuth and antimony salts. (Reserve for exam. by Table F.) If the solution be *neutral,* antimony sulphide—Sb_2S_3—might be here precipitated, *e.g.* from a solution of Na_3SbS_4, in water, sulphuretted hydrogen being evolved (see Table F). Aqueous solutions of certain double cyanides are also decomposed by hydrochloric acid; hydrocyanic acid being evolved, and insoluble cyanides precipitated. (See Analysis of Cyanides, page 146.)

If the solution be *alkaline, hydrochloric acid* may precipitate arsenious or stannic sulphides (As_2S_3, or SnS_2), as well as lead or silver chlorides. Note the *colour* of the precipitate.

Silicic acid might also be precipitated from an alkaline solution. This precipitate will be *gelatinous* (see Table M).

TABLE F.

EXAMINATION OF PRECIPITATE BY SULPHURETTED HYDROGEN.*

The Precipitate, after being *thoroughly* washed with water containing a few drops of sulphuretted hydrogen water, is warmed gently with ammonium sulphide and filtered.

A. Residue insoluble in Ammonium Sulphide.	B. Portion soluble in Ammonium Sulphide.
I Boil with dilute nitric acid.†	Acidulate with hydrochloric acid. Collect and wash precipitate. Treat with ammonium carbonate solution (see (*a*) page 114). Filter.
a. Insoluble portion. Dissolve in aqua regia, dilute, and add stannous chloride, white precipitate turning grey due to the passage of HgCl to Hg. MERCURY, as *mercuric salt.*	*a. Solution.* Acidulate with hydrochloric acid. Yellow precipitate = As_2S_3. ARSENIC. (Confirm by dry reactions, page 114.)
b. Solution. Add dilute sulphuric acid, and evaporate to small bulk, white residue insoluble in water = $PbSO_4$. LEAD.	*b. Insoluble part.* Dry : fuse with ammonium nitrate (see page 118). Fused mass treat with tartaric acid solution and filter.
Filter, add ammonia; white precipitate = $Bi(HO)_3$. BISMUTH. (Confirm by (*b*) and (*c*), page 112.)	1. Solution. Add hydrosulphuric acid gas. Orange precipitate = Sb_2S_3. ANTIMONY.§
Filter : a blue filtrate shows presence of COPPER. (Confirm by (*c*), page 113.)	2. *Residue.* Dry : test by match reaction for tin. Malleable metallic globules = Sn. TIN.
Acidulate this blue-coloured solution with hydrochloric acid, and treat it with hydrosulphuric acid gas ; collect precipitate and *wash well* :‡ boil in dilute sulphuric acid : filter. To filtrate add hydrosulphuric acid water. Yellow precipitate = CdS. CADMIUM.	

* If a strong nitric acid solution is treated with sulphuretted hydrogen, this gas is decomposed, sulphur being thrown down. Nitric acid must therefore be removed as far as possible by evaporation before adding the group reagent.

† Before adding nitric acid, make sure that the precipitate is *thoroughly* washed.

‡ If this washing be not properly conducted, part of the copper sulphide will be dissolved when sulphuric acid is added, and a brown-coloured precipitate again obtained on addition of sulphuretted hydrogen, even in presence of cadmium. Certain members of the next group yield precipitates with sulphuretted hydrogen. They are Fe_2O_3 and CrO_3; the precipitate is *white*, being merely sulphur; Fe_2O_3 is reduced to FeO, CrO_3 to Cr_2O_3, the *solution turning green.*

§ If this precipitate be *white*, it contains only sulphur formed by the decomposition of ammonium sulphide by hydrochloric acid.

TABLE G.

EXAMINATION OF PRECIPITATE BY AMMONIA IN PRESENCE OF AMMONIUM CHLORIDE.*

This precipitate may contain manganese in addition to the members of the group proper. It may also contain certain salts, as phosphates, oxalates, &c, which are held in solution by a free acid, but on the addition of Ammonia are precipitated unchanged. Two cases are thus presented.

A. Aluminium, Iron, Chromium, and Manganese only are present.

Precipitate is *well* washed, dissolved in nitric acid,† with addition of a few crystals of potassium chlorate, and caustic potash added until just neutral (no precipitate must be formed). When the solution has become perfectly cold, add excess of barium carbonate, and allow to stand for some time. Filter.

a. *Residue.* Wash, boil with caustic soda, filter, acidify filtrate with nitric acid and add ammonium carbonate; white precipitate=Al_2O_3. ALUMINIUM. (Confirm by dry reaction, page 120.)

The residue left on boiling with caustic soda is dried, fused with potassium nitrate, fused mass boiled with water, filtered, and to filtrate acetic acid in excess and lead acetate added. Yellow precipitate=$PbCrO_4$. CHROMIUM.

That part of the fused mass which is insoluble in water is dissolved in hydrochloric acid and potassium ferrocyanide added. Blue precipitate=Fe_3Cy_{12}. IRON.

b. *Filtrate.* Boil and precipitate excess of barium with sulphuric acid, filter, add caustic soda to filtrate. Brown precipitate=$Mn(HO)_2$. MANGANESE. (Confirm by dry reaction, page 125.)

B. Phosphates or Oxalates of the Alkaline Earths are also present.

Boil a part of the precipitate with excess of sodium carbonate, filter, acidulate with acetic acid, and add calcium chloride. White precipitate=CALCIUM OXALATE.

Oxalates are present. Group precipitate is dried and ignited (to decompose oxalic acid), residue dissolved in hydrochloric acid, ammonia added in excess and precipitated with ammonium sulphide. (The filtrate from this precipitate contains the earths formerly combined with oxalic acid. Test for them according to Table I.)

This precipitate, or in case of oxalates being absent, the original group precipitate, is dissolved in hydrochloric acid, and the solution boiled with potassium chlorate. In a small portion, iron is tested for with potassium ferrocyanide; in another small portion, phosphoric acid is tested for with ammonium molybdate. The remainder of the solution is neutralised with ammonia boiled with ammonium acetate (if no reddening of the liquid ensues, add two drops of ferric chloride) and filtered. (Test filtrate for manganese and alkaline earths.)

Precipitate is boiled with excess of caustic potash filtered, and filtrate tested for aluminium. The portion insoluble in caustic potash is fused with potassium nitrate and tested for chromium.

* Before adding these group reagents, expel *all* sulphuretted hydrogen from the liquid by continued boiling for 5 or 10 minutes, then add a little nitric acid and boil again to oxidise the ferrous salts which may remain.

† A small quantity of a white insoluble substance may remain. This is probably barium sulphate. When CrO_3 is reduced by sulphuretted hydrogen (see note ‡ to Table F), a sulphur acid, which is eventually oxidised to sulphuric acid, appears to be formed. This unites with barium, if traces of that metal are present in the solution.

M

TABLE H.

EXAMINATION OF PRECIPITATE BY AMMONIUM SULPHIDE.

Precipitate is washed with water containing a little Ammonium Sulphide, and then once with water alone.*
Treat the precipitate with *cold dilute* Hydrochloric Acid—filter

A. *Residue.*	B. *Filtrate.*
Test for cobalt in borax bead. (Table B.) Dissolve in hydrochloric acid (concd.) with addition of one crystal of potassium chlorate, evaporate nearly to dryness, dilute, neutralise with sodium carbonate, add potassium cyanide (free from cyanate), until the precipitate which at first forms redissolves, boil, add, while hot, a solution of mercuric oxide in mercuric cyanide, pale green precipitate=Ni(HO)$_2$, NICKEL (confirm by dry reaction, p. 122). Filter, evaporate filtrate to dryness, and test residue for cobalt by borax bead (see p. 124).	Boil so long as sulphuretted hydrogen is evolved (filter if necessary). Add caustic potash. Precipitate =Mn(HO)$_2$, changing to Mn$_2$O$_3$. MANGANESE: confirm by borax bead (Table B). Filter, and to filtrate add ammonium sulphide. White precipitate=ZnS, ZINC, confirm by flame test (p. 124).

* Should the filtrate and washings be very dark in colour, this is owing to dissolved nickel sulphide; in this case evaporate until excess of ammonium sulphide is expelled, then add dilute hydrochloric acid to acid reaction, and collect the black precipitate which forms along with the rest of the group precipitate.

TABLE I.

EXAMINATION OF PRECIPITATE BY AMMONIUM CARBONATE IN PRESENCE OF AMMONIUM CHLORIDE.*

Wash with hot water; dissolve in hydrochloric acid. Add ammonium sulphate solution, boil and filter.

A. *Precipitate*.

After washing, remove to test-tube, and boil with a mixture of two parts carbonate and one part sulphate of potassium; filter while hot. Over the residue in the filter pour hydrochloric acid.

Residue = BaSO$_4$.
BARIUM.
(Confirm by flame reaction, see p. 126.)

Solution.

Evaporate to dryness, and test residue in cold part of Bunsen flame; crimson-coloured flame, STRONTIUM.

B. *Filtrate*.

Neutralise with ammonia, and add ammonium oxalate. White precipitate = CaC$_2$O$_4$.
CALCIUM.

To a part of the filtrate from the *group precipitate* add sodium phosphate, and agitate briskly. White precipitate = Mg(NH$_4$)PO$_4$. MAGNESIUM.†

---

* The solution to which the group precipitant is added must be very gently warmed, not boiled.
† If a white *flocculent* precipitate forms, it may contain traces of aluminium phosphate or calcium phosphate.

M 2

TABLE K.

EXAMINATION FOR SODIUM, POTASSIUM, AND AMMONIUM.

A. *Ammonium.*

The original substance or solution is heated with caustic potash. Ammonia is thus evolved, recognisable by its smell, by its forming white fumes with hydrochloric acid, and by its action on red litmus paper. (See IV. (*d*), p. 130.)

B. *Potassium and Sodium.*

The solution in which magnesium has been detected is evaporated to dryness, and ignited to get rid of ammonium salts. The residue is dissolved in water, again evaporated to dryness, a few crystals of oxalic acid added, and the residue again ignited (the alkalies are thus converted into carbonates), water is added, the solution filtered, and the filtrate evaporated to dryness. Part of the residue is tested in the flame for sodium (p. 129); the remainder is dissolved in a few drops of water, and tested for potassium by means of platinum tetrachloride (p. 129).

If magnesium be absent, the residue, after ignition to expel ammonium salts, is at once tested for sodium and potassium.

TABLE L.

PRELIMINARY EXAMINATION FOR THE ACIDS.

When the base has been discovered, refer to the Table of Solubilities. If the substance is soluble in water, certain acids are excluded, because their compounds with the base present are insoluble in water. Some idea as to the acid present is thus obtained.

Heat a little of the substance with two or three times its volume of strong Sulphuric Acid.

A. No action ensues.

Silicic, boric, phosphoric, sulphuric, iodic, arsenic,* titanic acids.

B. Vapours are evolved.

a. Coloured Gases.

Hydriodic	Violet vapours, colouring starch paper blue.
Hydrobromic	Reddish vapour, colouring starch paper orange.
Chloric	Greenish yellow, explosive gas.
Hypochlorous	Yellow vapour; smell of chlorine.
Nitrous, and under certain circumstances nitric acid	Brownish yellow, irritating vapours.

b. Colourless Gases.

Acetic	Smell of vinegar.
Hydrofluoric	Fuming gas; corrodes glass.
Carbonic	Inodorous; gives turbidity with lime water.
Sulphurous.	Odour of burning sulphur.
Thiosulphuric.	" " "
Cyanic	Evolves carbon monoxide, which burns with lavender flame.
Ferrocyanic	" " "
Chromic	Evolves oxygen ; solution turns brown or green.
Oxalic	Evolves carbon monoxide and dioxide : former burns with lavender flame; latter gives turbidity with lime water.
Hydrosulphuric	Smell of rotten eggs ; blackens paper moistened with lead acetate solution.
Hydrochloric	Irritating odour ; gives white fumes with ammonia.

* This acid, or arsenious, if present, will have already been discovered among the bases ; so also will chromic acid. (See page 121.)

TABLE M.

GROUPING OF THE INORGANIC ACIDS (INCLUDING OXALIC ACID).

Divide the solution into three parts. Add to the *neutral solution* Barium Chloride.*

A. A precipitate forms.

a. *The precipitate is white* ; and
   α. SOLUBLE IN HYDROCHLORIC ACID. Phosphoric, carbonic, sulphurous, boric,† oxalic, hydrofluoric acids.
   β. INSOLUBLE IN HYDROCHLORIC ACID.‡ Sulphuric and hydrofluo-silicic acids.

b. *The precipitate is yellow* (and soluble in hydrochloric acid). Chromic acid.

c. *The precipitate is white* : it is decomposed by hydrochloric acid with separation of a gelatinous precipitate. Silicic acid. (Confirm by Table B, B 6.)

B. No precipitate forms.

To another portion of the neutral solution add silver nitrate.

C. A precipitate forms.

a. *The precipitate is white* : and
   α. SOLUBLE IN NITRIC ACID. Carbonic, sulphurous, thiosulphuric (soon turns black), boric, oxalic, and nitrous acids.
   β. INSOLUBLE IN NITRIC ACID. Hydrochloric, hydrobromic, cyanic, and hypochlorous acids.

b. *The precipitate is coloured—*
   Yellow . . . Soluble in nitric acid. Phosphoric acid.
   Light yellow . Insoluble in nitric acid. Hydriodic acid.
   Red brown . . Soluble in nitric acid. Chromic acid.
   Black . . . Insoluble in nitric acid. Hydrosulphuric acid.

D. No precipitate forms.

Test the original solution for nitric, chloric, and per-chloric acids. (See pp. 149 and 151.)

---

* If the solution be acid, neutralise with ammonia ; if alkaline, neutralise with nitric acid. Arsenious, arsenic, or chromic acid, if detected among the bases, must be removed by passing sulphuretted hydrogen through the *slightly acidulated* solution, filtering, boiling to remove excess of sulphuretted hydrogen and neutralising with ammonia. If carbonic acid has been detected by Table L, remove this by boiling after addition of nitric acid. If a member of Group I. of the bases has been proved present, add barium nitrate, instead of chloride.

† Borate and oxalate of barium, being soluble in solutions containing ammonium salts, would not be here precipitated if the solution has been neutralised with ammonia.

‡ As barium nitrate and chloride are nearly insoluble in concentrated hydrochloric acid, the acid employed here must be dilute. Sulphides, when heated with nitric acid, are oxidised to sulphates. If sulphides are suspected to be present, boil a part of the original substance in hydrochloric acid, and test the solution for sulphuric acid.

TABLE N.

FURTHER SEPARATION OF THE INORGANIC ACIDS.

A. Barium Chloride has produced a precipitate.

I. *The precipitate is white and insoluble in hydrochloric acid.* Dry the precipitate; heat a little in a test-tube, to which is adapted a bent tube dipping under water. If a gas is evolved ($SiFl_4$), the acid causing a gelatinous precipitate in the water (SiO_2), the acid present is HYDROFLUOSILICIC.

If this reaction does not take place, the acid present is SULPHURIC. Confirm by heating a little of the precipitate on the match in the upper reducing flame, by Table B, C4.

II. *The precipitate is white and soluble in hydrochloric acid.* Hydrofluoric, oxalic, carbonic, and sulphurous acids will have been detected by Table L. If these are absent, the precipitate must be owing to the presence of PHOSPHORIC or BORIC acid. Test part of the hydrochloric acid solution for boric acid, according to IV. (*d*), p. 135; the other part for phosphoric acid, according to III. (*a*), p. 137.

The other cases under this division require no further explanation.

B. Silver Nitrate has produced a precipitate.

I. *The precipitate is white and soluble in nitric acid.* All the acids here will have already been detected either by Table L, or by Div. A of this Table.

II. *The precipitate is white and insoluble in nitric acid.* Test a part of the original solution for hydriodic acid (because, although silver iodide is yellowish in colour, it might be here present *along with* the others).

Test for HYDROBROMIC and HYDROCHLORIC ACIDS according to XIII. and XIV. (*c*), p. 142 and 143.

Test another portion for HYDROCYANIC ACID by XVI. (*c*), p. 146.

A third portion is tested for HYPOCHLOROUS ACID by adding lead nitrate, a white precipitate changing to red, then to brown (from formation of lead oxide PbO_2), is produced.

Nitric acid will also have evolved the characteristically smelling hypochlorous acid (HClO) from the precipitate.

III. *The precipitate is coloured and soluble in nitric acid.* CHROMIC or PHOSPHORIC ACID.

Chromic acid will have been found among the bases. The colour of silver chromate is also characteristic.

If arsenic acid be present, this must be removed before testing for phosphoric acid (do this by means of sulphuretted hydrogen); then add to the acid solution ammonium molybdate, and heat gently. Filter off any precipitate, dissolve in ammonia, and apply test in III. (*a*), p. 135.

IV. *The precipitate is coloured and insoluble in nitric acid.*
a. *Yellow.* HYDRIODIC ACID; test a little of original solution by XIV. (*c*), p. 144.
b. *Black.* HYDROSULPHURIC ACID; already detected by Table L.

TABLE O.

EXAMINATION OF INSOLUBLE SUBSTANCES.

In this Table are included Silicates and Silica ; Alumina ; Sulphates of Barium, Strontium, and Lead ; Chloride, Bromide, and Iodide of Silver ; Fluoride of Calcium ; Tin, Antimony, and Chromium Oxides ; Lead Sulphide, Carbon and Sulphur.

Heat a portion of the substance in a dry tube.

A. It fuses.

(a) *Volatilises.* SULPHUR. Smell of sulphur dioxide.

(b) *Does not volatilise.* SILVER IODIDE, CHLORIDE, OR BROMIDE. (Match test, Table B, C 2.)

B. It is infusible.

(a). *After heating, it disappears.* CARBON.

(b). *It is white, unaltered by heating.*
Lead, barium, or strontium sulphate, calcium fluoride, silica or alumina.
1. Match-test for LEAD. (Table B, C₁.)
2. Bead of microcosmic salt in Bunsen flame for SILICA. (Table B, Bₐ.)
3. Heat in loop of platinum wire in flame ; moisten with cobaltous nitrate, and heat again. Blue mass=ALUMINA.
4. CALCIUM FLUORIDE. See p. 138, VI. (c).
5. BARIUM or STRONTIUM SULPHATE fused with sodium carbonate, gives on addition of hydrochloric acid, barium or strontium chloride, former giving green flame colouration, latter crimson.

(c) *It becomes darkened on heating, but original colour returns on cooling.*

(d) *It is coloured* (green), *and unaltered by heating.*

(e) *It is black.*

TIN or ANTIMONY OXIDE. (Match-test, Table B, C₂ and ₃.)

CHROMIC OXIDE (or chrome iron ore). (Table B, Bₐ.)

LEAD SULPHIDE. (Table B, C₁.)

Further tests may be applied by fusing the substance with 3 to 4 times its weight of a mixture of sodium and potassium carbonates (if any reducible metals, as lead or silver, are present, a porcelain crucible must be used) ; extract fused mass with water when cold, filter ; acidify filtrate with hydrochloric acid, evaporate to dryness, moisten with hydrochloric acid, add water, and filter ; test any insoluble substance for silica ; neutralise the filtrate with ammonia, and proceed according to Tables M and N.

The portion of the fused mass insoluble in water is dissolved in acid (hydrochloric, if lead and silver are absent), and tested for the bases.

TABLE P.

SEPARATION OF ORGANIC ACIDS.

The precipitate by calcium chloride, after standing 15 or 20 minutes, is filtered off.

A. *Precipitate.* (Calcium oxalate and tartrate.) Wash; add cold caustic potash solution.

Residue.

CALCIUM OXALATE. (Confirm by V. (*c*), p. 138.)

Solution.

CALCIUM TARTRATE. Reprecipitated on boiling.

B. *Filtrate.*

Boil for some time. CALCIUM CITRATE is precipitated, which is soluble in hydrochloric acid, but reprecipitated on addition of ammonia and boiling. Filter; add thrice the volume of alcohol, and filter.

Precipitate.

SUCCINIC ACID.

Filtrate.

Boil to expel alcohol. Neutralise with hydrochloric acid, and add ferric chloride. Buff-coloured precipitate.— BENZOIC ACID.

Another portion of the original solution is distilled with sulphuric acid, distillate neutralised with sodium carbonate, and evaporated to dryness. One part is tested for ACETIC ACID by V. (*b*), (*d*), p. 154; the other part is dissolved in water, and tested for FORMIC ACID by VI. (*b*), p. 154.

TANNIC and GALLIC ACIDS must be separately tested for in the original solution. With the first of these gelatin produces a *yellowish flocculent precipitate*; with the latter *no precipitate.* Ferric chloride in both solutions produces a bluish-black precipitate.

SECTION III.

DETECTION OF THE RARE ELEMENTS.

HITHERTO we have confined our attention to those elements or combinations of elements which are of common occurrence. There are, however, many of the so-called 'rare elements,' which are found very widely scattered among the mineral constituents of the earth; but, as they generally exist in only small quantities in any one mineral, there is occasionally some difficulty in detecting them.

DIVISION I.

Metals belonging to Group I. precipitated on addition of Hydrochloric Acid, **Thallium, Tungsten,** and **Niobium.**

1. THALLIUM.—The commonest source of thallium is *iron pyrites.*

One part of thallium in 10,000 parts of this mineral may be detected by the following process :—

After solution of the finely-powdered mineral in aqua regia, sulphuric acid is added, and the solution is evaporated until the nitric acid is driven off. On addition of sodium sulphite, the iron and thallium are reduced to protosalts; from this solution potassium iodide precipitates, after a little time, bright-yellow thallium iodide.

The black residue which remains when zinc is dissolved in sulphuric acid often contains thallium; this is dissolved in nitric acid, the solution evaporated with sulphuric acid, &c., and treated as above described.

II. TUNGSTEN, W : AND NIOBIUM, Nb.—Tungsten is generally found associated with iron and manganese, and sometimes with niobium in the mineral *wolfram.* To detect tungsten, the finely-powdered mineral is digested with strong hydrochloric acid, finally with addition of a little nitric acid. Iron and manganese are thus dissolved; the residual tungstic acid is dissolved in ammonia (if any white insoluble matter remains, remove it by filtration, *vide infra*), the solution is evaporated to dryness, and the residue ignited; pale-yellow tungstic oxide (WO_3) then remains. Heat a small portion of this with microcosmic salt in the reducing flame; a pure blue-coloured bead is obtained. The white portion, insoluble in ammonia, contains the niobic acid; it is tested in the microcosmic bead; in the oxidising flame it gives a colourless bead, and in the reducing flame a bluish-violet bead. On adding a trace of ferrous sulphate, the colour changes to blood-red.

Before testing for niobium, every trace of tungstic oxide must be removed, as it gives reactions very similar to those afforded by niobium oxide.

DIVISION II.

Metals belonging to Group II.

Palladium, Rhodium, Osmium, Ruthenium. (Div. I.)
Platinum, Iridium, Gold, Molybdenum, Selenium, Tellurium. (Div. II.)

PALLADIUM, RHODIUM, OSMIUM, RUTHENIUM, PLATINUM, AND IRIDIUM.—These metals are generally found associated together in the so-called *platinum ores.* Iridium is often present in certain ores of gold. By heating nearly to boiling, a quantity of the platinum ore, with five parts fuming hydrochloric and one part fuming nitric acid, in a small retort to which a well-cooled receiver is adapted, a distillate is obtained in which osmium (existing as osmic acid OsO_4) may be detected by the addition of potassium nitrite, potassium osmite, a red salt, being formed ; or by adding sodium sulphite when the solution turns violet, and after a time deposits blue osmium sulphite ($OsSO_3$).

The solution in the retort is separated from any black insoluble matter (consisting of osm-iridium), evaporated to dryness, the residue again evaporated with a little hydrochloric acid, and finally dissolved in water. On adding a few drops of this solution to a solution of sodium thiosulphate containing a little ammonia in a test-tube, a pale lemon-colour changing to brown, and eventually—on boiling —to black, is produced, if palladium is present.

To another portion of the solution concentrated by evaporation ammonium chloride is added ; platinum and iridium are thus precipitated. Wash with cold water, and in the filtrate precipitate palladium (as $PdCy_2$), by addition of mercuric cyanide; after removing this precipitate, sodium carbonate is added in excess, then hydrochloric acid, the solution evaporated to dryness, and the residue treated with alcohol. Any insoluble matter consists of sodium chlororhodiate.

Any ruthenium present in the solution may be detected, after the removal of the palladium, by boiling with a little hydrochloric acid, and adding ammonia to alkaline reaction. On now adding a solution of sodium thiosulphate and boiling, a fine red-purple colour (black by transmitted light) is produced in the liquid.

GOLD may be detected when existing in minute quantities by the following process : five or ten grams of the finely powdered mineral are shaken with alcoholic tincture of iodine, the insoluble matter is allowed to settle, a piece of Swedish filter paper is dipped into the solution, and incinerated after drying. If the ash be purple in colour, gold is present. To confirm the presence of gold, treat the ash with a few drops of aqua regia, evaporate to dryness at a gentle heat, and dissolve the residue in water ; pour this solution into a beaker which is set upon a sheet of white paper. A solution is now prepared by adding ferric chloride to stannous chloride until a permanent yellow colour is produced. This solution is diluted, a glass rod is dipped into it, then into the gold solution; a bluish-purple streak in the track of the rod confirms the presence of gold.

SELENIUM is sometimes found in specimens of native sulphur. Its presence may be detected by heating the powdered substance in potassium cyanide solution on the water bath for some hours (metals forming soluble compounds with potassium cyanide are assumed to be absent). On the addition of hydrochloric acid to this solution selenium only is precipitated. This precipitate, exposed on the prepared match to the reducing Bunsen flame, evolves an odour of decaying horse-radish.

The deposit which forms in the sulphuric acid chambers sometimes contains selenium. It is dried, deflagrated with potassium nitrate and carbonate in a red-hot crucible, the residue, which contains potassium selenate, is digested with hydrochloric acid, evaporated to a small bulk, saturated with sulphur dioxide, and boiled. Red flakes of selenium are precipitated.

MOLYBDENUM occurs in *molybdenite* as sulphide, and in *wulfenite*, combined with lead. The mineral is treated with dilute hydrochloric acid until any iron and zinc present are dissolved, the acid is poured off, and the residue treated with strong hydrochloric acid, and finally evaporated to dryness. The mass is digested with ammonia, filtered, and the ammoniacal solution evaporated down. On cooling, crystals of ammonium molybdate are deposited, which may be dissolved in nitric acid. This solution, on addition of sodium phosphate, gives a yellow precipitate.

TELLURIUM occurs native, associated with bismuth, lead, gold, and silver; sometimes also with selenium.

The mineral, in fine powder, is fused, with ten times its weight of potassium cyanide, in a long-necked flask through which a stream of hydrogen is passed. Potassium telluride is formed. The residue is dissolved in water, and a current of air passed through the solution, whereby the tellurium is precipitated in metallic scales.

These scales, heated in the charcoal splinter, are volatilised, forming a white deposit of tellurous oxide on the charcoal. A slightly acid smell is alone produced, very different from the strong odour produced by selenium compounds (*vide supra*).

DIVISION III.

Metals belonging to Group III.

Zirconium, Cerium, Lanthanum, Didymium, Titanium.

ZIRCONIUM usually occurs in nature in *zircons* or *jargons*. These are decomposed by fusing with about five times their weight of pure sodium carbonate, and adding a small piece of pure sodium hydrate after fusion is nearly complete. On treating the fused mass with cold water the silica is removed. The insoluble portion is dissolved in hydrochloric acid, evaporated to dryness, and the residue treated with water acidulated with hydrochloric acid ; the solution now contains nearly pure zirconium chloride. From this solution oxalic acid throws down a bulky precipitate of zirconium oxalate, soluble in hydrochloric acid.

Ammonium and sodium carbonates precipitate zirconium carbonate, soluble in large excess of the precipitants.

Potassium sulphate added to a hot solution of zirconium salts precipitates zirconium sulphate, insoluble in water and in hydrochloric acid.

The three metals, Cerium, Lanthanum, and Didymium, usually occur together in the mineral *cerite*. The finely-powdered mineral is boiled for an hour or so with strong hydrochloric acid ; the metallic oxides are thus dissolved ;

from the filtered solution, on addition of ammonia in excess, everything but the lime is precipitated. This precipitate is redissolved in nitric acid, the solution diluted with its own volume of water, a small quantity of pure peroxide of lead added, and the liquid boiled ; the cerium is thus oxidised, its presence being shown by the yellow or orange colour produced in the liquid. This liquid is evaporated to dryness, heated so as to expel a portion of the acid, and the residue treated with boiling water, acidulated with nitric acid ; lanthanum and didymium, under these circumstances, are alone dissolved. By evaporating this solution, excess of lead having been removed by sulphuretted hydrogen, heating the residue for a few minutes to 400° or 500°, and subsequently adding hot water and filtering, the nitrate of lanthanum is found in solution ; in this solution caustic potash produces a precipitate of lanthanum hydrate, which yields a perfectly colourless bead when tested in the Bunsen flame with borax (an amethyst colour is due to traces of didymium).

The portion insoluble in hot water consists of didymium subnitrate; heated with microcosmic salt in the reducing flame an amethyst bead is obtained.

TITANIUM generally exists in the form of oxide TiO_2 ; it is found in certain iron ores, in many silicates, and in some iron slags. To detect its presence, fuse a portion of the substance in fine powder, with about eight times its weight of potassium bisulphate, triturate the fused mass with cold water until all is dissolved excepting silica, filter, add a little sodium thiosulphate to the filtrate, and boil ; a yellowish-white precipitate (TiO_2) shows the presence of titanium ; confirm by testing with microcosmic salt in the oxidising flame. A colourless glass is produced ; but in the reducing flame the bead is yellow while hot, turning violet on cooling.

DIVISION IV.

Metals belonging to Group IV.

Uranium and Indium.

URANIUM occurs in a few minerals, most abundantly in *pitchblende* and *uranite.* To detect its presence in these substances, where it is usually mixed with iron, &c., advantage is taken of the fact that its carbonate is soluble in excess of ammonium or sodium carbonate, in which respect it differs from iron. To prove the presence of uranium, dissolve a portion of the mineral by boiling with a small quantity of nitric acid, dilute with water, add a tolerably large excess of a solution of sodium carbonate, and boil ; filter from insoluble carbonates of iron, calcium, &c., and to the filtrate add caustic potash ; a yellow precipitate of uranic oxide (U_2O_3) is found. In the borax bead a yellow glass, becoming green on cooling, is produced by heating in the oxidising flame ; in the reducing flame the bead is green both when hot and cold.

INDIUM is found in some specimens of zinc. The sample is almost entirely dissolved in hydrochloric acid. After about twenty-four hours the clear liquid is poured off, the residue is treated with a few drops of sulphuric acid, washed with hot water until it has no longer an acid reaction, nitric acid is added, and a few drops of sulphuric acid, the liquid evaporated nearly to dryness, water added, and the solution separated by filtration. Ammonia in large excess is added to the filtrate, the ensuing precipitate collected, well washed, dissolved in a very few drops of hydrochloric acid, acid sodium sulphite added in excess, and the solution boiled. A salt, having the composition $2In_2O_3$, $3SO_2 + 8H_2O$, separates out. On the charcoal splinter white ductile metallic beads are obtained. On the asbestos thread a violet-blue colouration is imparted to the flame.

Vanadium, Lithium, Cæsium, Rubidium.

VANADIUM occurs in many lead ores, also in small quantities in several other minerals, and in certain sandstone beds containing copper. To detect vanadium decompose the mineral with strong nitric acid, add sulphuric acid, and filter from insoluble lead sulphate (wash this precipitate with alcohol); the filtrate contains vanadium pentoxide (V_2O_5). On addition of zinc, the solution goes through a series of changes in colour, ultimately a violet solution being obtained, which, on standing in the air, turns brown by oxidation, or if excess of acid be present becomes green.

The three alkaline metals, Lithium, Cæsium, and Rubidium, occur in the mineral *lepidolite* and in certain mineral waters. They seem to be very widely diffused, occurring, however, generally in but minute quantities. To detect them in lepidolite, fuse the powdered mineral, pour it into cold water, pulverise and wash, then add about twice its weight of hydrochloric acid, and boil for some time; filter, add a little nitric acid, and boil to peroxidise iron; *dilute the solution* and add sodium carbonate, filter and boil down the filtrate; now add hydrochloric acid in excess, and platinum tetrachloride; cæsium, and rubidium are thus precipitated as double chlorides Cs_2PtCl_6 and Rb_2PtCl_6. In the filtrate, after removing excess of platinum by means of sulphuretted hydrogen, and concentration of the liquid, sodium carbonate precipitates lithium carbonate (confirm by testing in Bunsen flame or spectroscope).

To detect lithium in mineral waters evaporate part of the water to a small bulk, add baryta water, and, on cooling, ammonium carbonate, and filter; add sodium phosphate to the filtrate, evaporate to dryness, and treat the residue with a very small quantity of water. Lithium phosphate remains behind, and may be tested in the spectroscope.

To detect these 'rare metals' in the ordinary course of analysis the following method may be pursued :—

The precipitate of Group I. is boiled in water. Lead and thallium chlorides are thus dissolved ; filter, and to the filtrate add dilute sulphuric acid. To the filtrate from the precipitated lead sulphate add potassium iodide ; a yellow precipitate (TlI) indicates thallium (confirm by the spectroscope). That portion of the group precipitate which is insoluble in boiling water is treated with ammonia to remove the silver chloride from the residue. The mercury is removed by treatment with nitric acid. The insoluble part is now tested in the borax bead for tungsten (see p. 171).

That portion of the precipitate by sulphuretted hydrogen which is insoluble in ammonium sulphide may contain the sulphides of the rare metals, palladium, rhodium, osmium, and ruthenium.

The precipitate is to be fused with caustic potash and potassium chlorate, the fused mass treated with water, and filtered. Ruthenium is tested for in a portion of the filtrate by boiling with a little dilute hydrochloric acid, adding ammonia in slight excess, and then boiling with sodium thiosulphate solution ; a purple-red colour indicates ruthenium. Another portion of the filtrate is distilled after addition of nitric acid, and osmium tested for in the distillate according to test on p. 172.

That part of the fused mass which was insoluble in water is to be dried, gently ignited in a stream of hydrogen, the reduced metals very gently heated with dilute nitric acid, whereby copper, lead, bismuth, and cadmium are dissolved, and the insoluble portion treated with aqua regia. The residue contains any rhodium which may be present. Palladium may be detected in the aqua regia solution by the reaction with sodium thiosulphate (see p. 172). Mercury must be tested for in another portion of the original precipitate by sulphuretted hydrogen.

The precipitate resulting from the addition of acid to the

ammonium sulphide solution (containing the members of Division II.) may contain the sulphides of platinum, iridium, gold, molybdenum, selenium, and tellurium (in addition to those of arsenic, antimony, and tin).

The precipitate is dried, mixed with one part dry sodium carbonate and one part sodium nitrate, and the mixture projected into a porcelain crucible containing two parts sodium nitrate in a state of fusion. The fused mass, when cold, is triturated with cold water, and the aqueous solution separated by filtration. The filtrate contains (in addition to arsenic) the tellurium, selenium, and molybdenum, while antimony, tin, gold, platinum, and iridium remain in the insoluble part.

I. A portion of the aqueous solution is tested for arsenic according to V. (*c*) and (*d*), p. 114.

To another portion, acidified with hydrochloric acid, potassium sulphocyanide is added, and a rod of pure zinc immersed in the liquid; if **molybdenum** be present, the solution becomes crimson, owing to formation of molybdenum sulphocyanide. The remaining portion of the solution is evaporated to dryness, and the residue fused with potassium cyanide in a stream of hydrogen. The fused mass is then dissolved in water, and a current of air passed through this solution. The whole of the **tellurium** is thus precipitated in the metallic state.

On adding hydrochloric acid and stannous chloride to the filtered solution a red precipitate of **selenium** is produced, if that element is present.

II. That portion of the fused mass which is insoluble in water is now to be tested for antimony, tin, gold, platinum, and iridium. Heat it very gently with a saturated solution of tartaric acid, filter, and test the filtrate with sulphuretted hydrogen for antimony. Transfer the residue to a platinum capsule, heat with hydrochloric acid, and add a small piece of pure zinc. If tin be present it is reduced to the metallic state. Cautiously pour off the liquid, wash the residue thoroughly, and again warm with hydrochloric acid. Test the solution thus obtained for tin (a greyish white precipitate with mercuric chloride), remove the part insoluble in hydrochloric acid to a porcelain dish, and dissolve in aqua regia.

Test one portion of this solution for gold by making use of the reaction with stannous and stannic chlorides (p. 173).

If platinum is present another portion will give, on addition of potassium chloride, a yellow crystalline precipitate of double potassium-platinum chloride ($2KCl.PtCl_4$), while a third portion, treated with caustic potash, turns deeper red in colour, and eventually, after exposure to the air, becomes azure blue, if iridium is present.

The group precipitate obtained by adding ammonium chloride and ammonia, may contain zirconium, titanium, cerium, lanthanum, and didymium.

The precipitate is dried, ignited, and fused for some time in a platinum crucible with potassium-hydrogen sulphate. On boiling the fused mass with water iron remains in the insoluble portion; the presence of this element may be recognised by dissolving in hydrochloric acid, and applying the tests for iron. The aqueous solution is diluted and boiled for some time, carbon dioxide being led through the liquid; if a precipitate forms filter it off, and test it for titanium in the borax bead (p. 175). To the filtrate add a strong caustic potash solution, filter and boil the liquid, test any precipitate which forms for chromium, and the filtrate therefrom test for aluminium. Dissolve the precipitate formed by caustic potash in hydrochloric acid, and remove the excess of acid by evaporation (if the precipitate was reddish in colour add barium carbonate and allow to stand for several hours; iron is then removed; filter, and to the filtrate add sulphuric acid to remove excess of barium, neutralise exactly with potash, and proceed as follows):—add a few crystals of potassium sulphate, boil, and allow the fluid to stand for some hours; now filter and wash with a dilute potassium sulphate solution, remove the precipitate to a test tube, and boil repeatedly with water and a little hydrochloric acid, the oxides of cerium, lanthanum and didymium are thus dissolved, while the potassium-zirconium sulphate remains undissolved.

The cerium, lanthanum and didymium may be separated by the process given on pp. 174, 175.

Before treating the precipitate of Group IV. with dilute hydrochloric acid, warm it very gently with a strong ammonium carbonate solution, filter, and test the filtrate for uranium according to the method given on p. 176.

The filtrate from Group IV. may contain vanadium and also traces of tungsten, unprecipitated by hydrochloric acid. To detect these, acidify the liquid with hydrochloric acid, filter and fuse the sulphides with sodium carbonate and potassium nitrate, dissolve the fused mass in water (any residue may consist of nickel oxide), and in one portion of the solution test for vanadium by adding solid ammonium chloride, a white precipitate of ammonium vanadate is obtained if vanadium is present : to the rest of the solution add hydrochloric acid and a piece of zinc, if the liquid becomes blue tungsten is present.

SECTION IV.

DETECTION OF POISONS.

A. *INORGANIC POISONS.*

(Phosphorus, Arsenic, Antimony, Mercury, Lead, Copper, and Zinc.)

THE general method to be employed when all poisons, inorganic and organic, are to be sought for, is as follows :—

Acidulate the mass, distil, and test the distillate for hydrocyanic acid (XV. (*c.*), p. 146).

Test the residue for phosphorus by method A or B
(*infra*). The residue is tested for alkaloids (see p. 201).
All the residues, extracts, &c., are then united, alcohol and
ether removed by evaporation, and the mass examined for
the metallic poisons (see p. 183).

I. PHOSPHORUS.—First, note whether the substance
under examination, when thoroughly stirred up, smells of
phosphorus, or is luminous in the dark. A slip of paper
moistened with silver nitrate is suspended in the neck of a
flask containing a little of the suspected matter, the flask
being heated to 30° or 40°. If the paper is not blackened
no unoxidised phosphorus is present, and, consequently,
any further examination may be dispensed with. If, however, the paper is turned black, it is absolutely necessary to
apply further tests before deciding that phosphorus is actually present.

The following processes may be employed :—

A. The suspected matter, mixed with water and a little
sulphuric acid, is boiled in the dark in a flask connected
with a Liebig's condenser. If phosphorus is present—even
in quantities so small as 1 part in 100,000—a luminous ring
appears in the upper part of the inner tube of the condenser.
Certain substances—such as alcohol, ether, oil of turpentine,
&c.—prevent the luminosity of phosphorus, so that, when
such bodies are present, this method fails. Easily volatilised
compounds—as alcohol and ether—soon, however, distil
over, and the luminosity then appears.

Small globules of phosphorus are found at the bottom of
the receiver. These may be collected, dried, and their
luminosity in the dark observed.

B. The suspected matter, along with water and dilute
sulphuric acid, is warmed gently for several hours in a flask
through which a slow stream of washed carbon dioxide is
passed, and with which is connected a U-tube, containing
silver nitrate solution. If a black precipitate (silver phos-

phide) forms in this solution, it is to be collected and examined as follows, making use of the apparatus shown in fig. 52 :—

a is a hydrogen generator ; *b* contains pumice-stone, moistened with strong caustic potash ; *c* is a *platinum* jet (kept cool by means of moistened wool) ; *d* and *e* are clips. Hydrogen is generated in *a*, the current of the gas being so regulated that a small, steadily-burning flame is obtained at *c*; if this flame shows no green inner cone, and

FIG. 52.

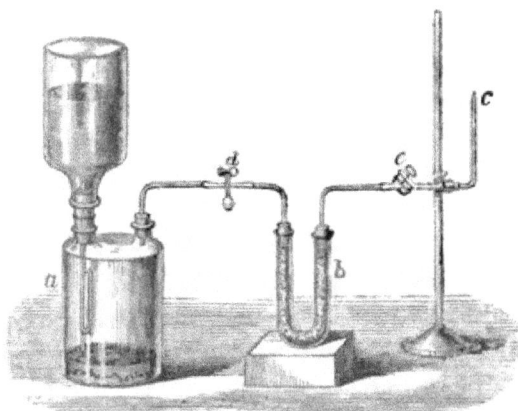

no emerald colouration when caused to impinge upon porcelain, we may conclude that the gas is free from phosphoretted hydrogen. The precipitate is now to be washed into *a* (through the upper flask), and the flame at *c* noticed. If even a trace of silver phosphide existed in the precipitate, a green cone becomes visible in the centre of the flame, which also shows an emerald colouration when it impinges on porcelain. This method is applicable in the presence of substances which hinder the reaction of method A.

II. METALLIC POISONS.—*Arsenic, Antimony, Mercury, Lead, Copper and Zinc.*—The viscera to be examined are placed in a large porcelain dish, there turned inside out, and carefully examined for small white grains of arsenious oxide.

As this substance is but slightly soluble in water, part of it not unfrequently remains in the undissolved state imbedded in the coats of the stomach, &c. If any such particles are found, they must be carefully dried by pressing between folds of blotting paper, and reserved for examination. After mixing all the contents of the dish together, one-third is set aside, and the remaining two-thirds mixed with distilled water, thoroughly stirring up the mass with a glass rod.

After standing a few minutes, pour the fluid into another dish ; repeat this operation several times, and finally carefully feel the residue in the dish with a glass rod to make sure that no small gritty particles remain in it. If any such are found, remove them, dry them, and reserve them for examination (see p. 194).

If time or the condition of the substance to be examined permit, the semi-liquid mass may now be submitted to dialysis. The dialyser consists of a round sheet of parchment paper, 30–35cm. diam., spread over a gutta percha hoop, 5cm. deep and 20–25cm. diam. The paper being previously wetted, another hoop is slipped over the first, so as to secure the paper. When the upper surface of the paper is wetted with a sponge, no drops should show themselves on the under surface. The liquid, after twenty-four-hours' digestion at 32°, is poured into this dialyser, covering the paper to the depth of 12mm. or so, and the whole is floated on water in another dish (see fig. 53). After twenty-four hours the greater portion of the dissolved arsenious oxide or other metal will have passed into the outer water. The dialysate is to be concentrated by evaporation on the water-bath, acidulated with hydrochloric acid, and treated with sulphuretted hydrogen, as directed on p. 185. The principal advantage of this method is that the poisons are obtained in a pure state in the dialysate, separated from colloidal organic matter. If no poison be detected, test the residue on the dialyser as directed in the following section :—

Care must of course be taken to insure the purity af the reagents used in all these operations.

The semi-liquid mass on the dialyser, or, if dialysis has not been resorted to, the original substance, is heated on the water-bath. A quantity of pure hydrochloric acid (Sp. Gr. 1·12), about equal in weight to the amount of solid substance, is added. together with distilled water, if necessary, sufficient to form a thin paste. Continue to heat the mixture on the water bath, and add potassium chlorate, in quantities of about two grammes at a time, until a light yellow · homogeneous fluid is obtained. The water lost by evaporation is to be replaced from time to time.

FIG. 53.

After a further addition of potassium chlorate, and when the liquid is cool, filter it, wash the residue well with hot water, collecting the washings separately from the filtrate. The filtrate is heated on the water-bath until it no longer smells of chlorine (evaporated water being replaced). The washings are evaporated to about 100 cc., and added to the filtrate. The liquid is now transferred to a flask, heated to about 70°, and a slow current of washed sulphuretted hydrogen passed through it for about twelve hours; after which the flask is covered over, and allowed to stand in a moderately warm place for twelve hours or so ; if the liquid now no longer smells of sulphuretted hydrogen, it must be again treated with the gas. Collect the precipitate which may have formed on a filter, and wash it (at first with water containing sulphuretted hydrogen) until the washings are free from chlorine. If no precipitate has been produced by the sulphuretted hydrogen, the liquid should be concentrated by evaporation at a gentle heat, and the gas again passed through it.

The precipitate may contain the sulphides of arsenic, antimony, mercury, copper, or lead, in addition to a considerable quantity of organic matter.

The filtrate, which may contain zinc, is concentrated by evaporation, transferred to a flask, ammonia and ammonium sulphide added, the flask corked up, and set aside for further examination (see p. 196). The precipitate is now to be dried in a small dish on the water-bath; pure fuming nitric acid (which must be free from chlorine) is then added, so as to completely moisten the mass, which is then evaporated to dryness on the water-bath. The residue is moistened with pure strong sulphuric acid, and heated first on the water, then in the air-bath, at about 170°, until the mass is friable. If this state is not at once obtained, a few cuttings of Swedish filter paper are added, and the heating continued. Now warm the residue with 1 part of hydrochloric acid and 8 parts water; filter, wash with hot water, and reserve the residue for examination (Residue A).

The filtrate must now be again saturated with sulphuretted hydrogen gas, attending to the directions already given, and the ensuing precipitate collected and washed. This precipitate is now perfectly free from organic matter.

Note the colour of the Precipitate.

If it is light yellow, probably arsenic only is present. Shake up a small portion with ammonia in a test-tube; if it dissolves entirely (*infra*) arsenic is the only metal present. If, however, it does not entirely dissolve, or if the precipitate is not yellow, you must test for other metals besides arsenic.

If arsenic is the only metal present, treat the precipitate on the filter with ammonia, whereby everything is dissolved (except perhaps a little sulphur). Evaporate the ammoniacal solution to dryness on the water-bath, and reserve it for subsequent examination. (B. *infra.*)

If other metals than arsenic are present, spread out the filter with the precipitate in a porcelain dish ; add a little ammonia and ammonium sulphide, and then wash the solid matter off the filter with the smallest possible quantity of water ; remove the filter, and gently heat the residue in the basin. Filter, *wash*, and reserve the residue for examination (Residue C., p. 195). The filtrate and washings are now evaporated to dryness, pure fuming nitric acid added, and the mass gently heated ; when dry, add a little caustic soda, which neutralises the acid without effervescence ; add now a mixture of 1 part carbonate and 2 parts nitrate of sodium, free from chlorine ; evaporate to dryness, and heat gradually until the mixture fuses to a colourless liquid. When cold treat the mass with cold water, and filter. Arsenic thus goes into solution as sodium arsenate, while antimony remains on the filter as sodium antimonate ; filter ; set aside the insoluble portion (D. see p. 196) ; add pure dilute sulphuric acid to the filtrate, and evaporate in a porcelain dish, adding a little more sulphuric acid towards the end of the operation. If arsenic was the only metal present, the residue B (*vide ante*) is treated with fuming nitric acid, and the resultant liquid evaporated down with sulphuric acid until all nitric or nitrous acid is expelled. Two methods may be employed for the further examination of this liquid.

1. *Marsh's Process.*—This method is based upon the fact that arseniuretted hydrogen (AsH_3) is formed when nascent hydrogen acts on a soluble arsenical compound, and that this gas, arseniuretted hydrogen, is readily decomposed by heat, yielding metallic arsenic.

A very convenient and easily-constructed form of apparatus is shown in fig. 54. It consists of a small, wide-

mouthed flask, *a*, capable of holding about 300 cc., fitted
with a good caoutchouc cork,* which is perforated with two
holes; through one of these passes a funnel tube, and
through the other a glass tube, bent at right angles, and
passing through a small cork into a small bulb-tube containing lumps of perfectly dry calcium chloride. To the
other end of this latter tube a piece of hard, combustion
tubing is adapted, by means of caoutchouc tubing. This
hard tubing, which should be about 1 cm. in diameter, is

FIG. 54.

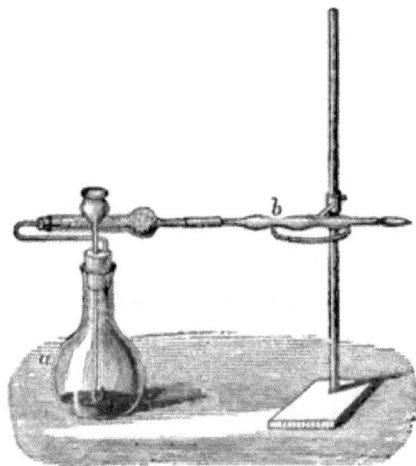

narrowed by drawing it out in the flame at intervals of about
2 cm., so as to present the form of a narrow tube, with
3 or 4 bulbs or wider portions on it (fig. 54, *b*).

The end of this tube is made slightly rough with a file,
dipped into a solution of platinum tetrachloride, and heated ;
a thin coating of platinum is thus formed on the surface of the

* If red corks or tubing be employed they must be boiled in caustic
soda and washed with water before being used, as they contain antimony, and generally a little arsenic sulphide.

tube, whicn prevents the glass fusing when the issuing hydrogen is ignited.

The hydrogen is most conveniently prepared by the action of dilute sulphuric acid upon zinc. For this purpose a quantity of granulated zinc is placed in the flask, and water added ; the whole apparatus is then put together, and a little strong sulphuric acid poured into the flask by means of the funnel-tube. The hydrogen thus generated is dried by passing over the calcium chloride. After a little time, and when a sample of the gas tested in the usual manner (see Lesson II.) is shown to be free from air, the low flame of a Bunsen lamp is applied to the first bulb of the reduction tube, and the evolution of hydrogen maintained for about fifteen minutes. If after this time not the least trace of a metallic-looking mirror is formed in the tube beyond the heated portion, we may conclude that the materials are free from arsenic.

We are now ready to begin the actual examination of the liquid supposed to contain arsenic. It will generally be necessary to disconnect the apparatus, pour off the acid liquid from the zinc, add a little more water, and then a few drops of strong sulphuric acid. After ascertaining that the issuing hydrogen is free from air, light it, and now apply a flame to the first bulb of the reduction tube ; when this is strongly heated, pour about one-half of the suspected liquid. prepared as directed (p. 187) gradually down the funnel tube ; the evolution of gas increases, and if arsenic is present a metallic mirror is formed between the heated portion and the outlet of the reduction tube. If the evolution of arseniuretted hydrogen is considerable, part of it will probably escape decomposition by the heat applied ; the flame at the end of the tube will change to a peculiar bluish-white tint ; at the same time white fumes of arsenious oxide will appear immediately above this flame. Having obtained two mirrors in the reduction tube, remove the lamp, and bring a piece of cold porcelain, *e.g.* a small basin or crucible lid, close to the end of the tube, so that the flame of the burning gas strikes against it. A brownish-black lustrous spot is produced on the porcelain.

Now extinguish the flame ; adapt a bent tube to the end of the delivery tube, and pass the gas into a solution of silver nitrate in a small test-glass. The liquid in the glass gradually blackens, metallic silver being deposited, while the solution contains arsenious acid ($6 AgNO_3 + AsH_3 + 3H_2O = 6 HNO_3 + 6 Ag + AsH_3O_3$).

The following precautions must be attended to in using these tests :—

1. A small quantity only of the acid liquid is to be poured at a time into the generating flask, otherwise the reaction will get very violent, and the contents of the flask will probably froth over.

2. The porcelain dish used for obtaining spots must be clean, dry, and at the temperature of the air. It must not be held for any length of time in the flame, but merely brought down upon the flame and withdrawn, otherwise the porcelain will become so heated that the arsenic spot will be vapourised.

3. Conduct the entire operation in a place where there is sufficient draught to carry off all arsenical fumes, as these are extremely poisonous.

If the process described for preparing the liquid to be tested for arsenic has been carefully performed, the presence of antimony in this liquid is almost impossible. As antimoniuretted hydrogen, however, gives reactions very similar to those of arseniuretted hydrogen, we must make sure that the mirror, spots, &c., are really due to the presence of arsenic.

(*a*) The mirror on the reduction tube, if caused by arsenic, is all deposited *after* the point at which the flame was applied, *i.e.* between the heated point and the outlet of the reduction tube.

Antimony being less easily vapourised than arsenic, is deposited *on both sides* of the heated spot. The arsenic mirror is lustrous, brownish-black in colour, and when in thin layers, and viewed against a sheet of white paper, is transparent. The antimony mirror when thus viewed appears as a brownish-black dust. Near the heated spot it is partly fused to small globules, and is almost silver-white in colour. That part of the tube containing the mirror is now to be cut off with a file, and reserved for examination.

(*b*) The spots on porcelain, if caused by arsenic, are lustrous ; if caused by antimony, they have a velvet-like appearance, and are nearly black in colour.

If the arsenic spots are touched with a drop of sodium hypochlorite, free from chlorine, they are dissolved, while antimony spots are unaffected by this reagent. If the spots are touched with a drop of yellow ammonium sulphide, and heated gently over a flame, they are dissolved. On now evaporating off the excess of ammonium sulphide by gently heating and blowing, a yellow residue remains if the spot was arsenic ; if antimony, the residue is orange. (A very good way of applying this reaction is to use a small wash-bottle, the blowing tube of which passes into a little strong ammonium sulphide, while the exit-tube is cut off immediately below the cork ; by this means a current of ammonium sulphide gas may be directed upon the spot; if arsenic, a yellow ring of arsenious sulphide (As_2S_3) is formed at the outer margin of the spot ; if antimony, this ring is orange-red in colour ; these rings disappear on continuing the passage of ammonium sulphide vapour over them ; but reappear on gently breathing upon them.)

(*c*) If the gas was arseniuretted hydrogen, the silver nitrate solution through which the gas has been passed contains metallic silver and arsenious acid (see p. 190), along with excess of silver nitrate. On filtering from the silver precipitate, and carefully neutralising the filtrate with ammonia (adding, if necesary, a little more silver nitrate), a yellow precipitate of silver arsenite (Ag_3AsO_3) is obtained. If the gas was antimoniuretted hydrogen, the precipitate in the silver nitrate solution will contain all the antimony combined with silver ($3AgNO_3 + SbH_3 = 3HNO_3 + Ag_3Sb$). On filtering, therefore, and adding ammonia, no precipitate will be obtained; but if the precipitate is washed with hot water, then boiled with a concentrated solution of tartaric acid, and the insoluble part removed by filtration, an orange precipitate of antimonious sulphide is obtained, on acidulating the filtrate with hydrochloric acid and passing sulphuretted hydrogen through it.

The part of the combustion-tube on which the mirrors are formed is cut off, and one-half of it sealed up in a small tube, and preserved as a *corpus delicti*, the other half is placed in a clean, dry, wide test-tube, and gently heated; the arsenic volatilises, is oxidised, and deposited as arsenious oxide on the wider tube, in the form of a white crystalline ring. The little bit of combustion-tube is thrown out from the test-tube, the arsenious oxide dissolved in water, this solution divided into two parts, and tested (1) with ammonio-silver nitrate (2) with ammonio-cupric sulphate (see p. 114 (*b*) and (*c*)). These tests taken together furnish conclusive evidence of the presence of arsenic.

2.—*Method of Fresenius and Von Babo.*—This process depends upon the fact that arsenious sulphide, when fused with potassium cyanide and sodium carbonate, is readily reduced, metallic arsenic being formed. The delicacy of the reaction is increased by heating the mixture in a slow current of carbon dioxide.

Fig. 55 shows the arrangement of the apparatus.

A is an apparatus for producing the carbon dioxide; B is a bottle containing strong sulphuric acid, which dries the carbon dioxide; C is a piece of hard glass tube, about 7 mm. bore and 30 cm. long.

The liquid obtained, as described on p. 187, is diluted with water and saturated with sulphuretted hydrogen gas, with attention to the directions already given. The resulting precipitate is washed, and excess of sulphur removed by treatment with carbon disulphide ; the residue, after being again washed, is dissolved in ammonia, and the solution evaporated to dryness.

About one-half of the dry arsenious sulphide thus obtained is mixed with ten times its weight of a mixture of 3 parts perfectly dry sodium carbonate and 1 part dry potassium cyanide. The powder is placed in a little piece of stout paper with turned-up edges, which is pushed into the

FIG. 55.

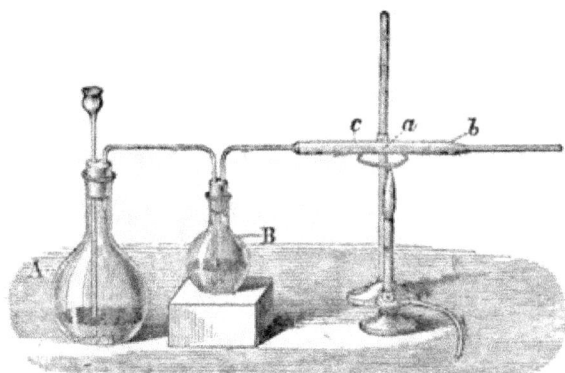

reduction-tube to *a* ; the tube is now turned half round on its axis, so that the powder falls out of the paper, which is then withdrawn. Pass a slow current of carbon dioxide through the tube, heating it gently along its whole length until *perfectly dry*. The gas should now be passing through B at the rate of one bubble per second ; heat the tube at *b* to redness, and apply the flame of a second lamp to the mixture, proceeding from the end nearest B to the point at which the mixture is situated. The arsenic is gradually expelled and condenses, chiefly at *b* ; the very small quantity which escapes may be recognised by its garlic odour. By advancing the second lamp slowly towards the first, all the arsenic is driven into the narrow portion of the tube.

The tests mentioned on page 192 may be applied to this mirror. By this process no mirror is obtained from any antimony compound.

If in the preliminary examination of the animal matter small pieces of solid matter, which appear like arsenious oxide, have been obtained (page 184), they may be examined in the following way :—

A small particle is introduced into the narrow portion of a reduction-tube (fig. 56), and a thin splinter of freshly ignited charcoal is dropped upon the top of it. That part of the tube where the charcoal is situated is heated to redness over a very low flame, and the tube gradually moved until the supposed arsenious oxide is also heated. If it vapourises, and forms a brownish-black lustrous ring above the charcoal, the substance is, in all probability, arsenious oxide. To make certain, heat the arsenical mirror ; it vapourises, is oxidised, and again condenses on the wider portion of the tube, in the form of a white crystalline sublimate. This is dissolved in water, and tested as directed on p. 192.

Fig. 56.

Another fragment of the white substance, supported on asbestos, is brought into the reducing Bunsen flame, and metallic films obtained on porcelain (p. 98) ; these are tested with sodium hypochlorite and ammonium sulphide, as already directed.

The several residues which you have set aside during the process must now be examined.

1. Residue A is to be examined for lead, mercury, and antimony ; it is dissolved in aqua regia, any insoluble matter being removed by filtration ; the solution is boiled down to a small bulk, hydrochloric acid added, and the liquid again evaporated ; it is then diluted with water, and saturated with sulphuretted hydrogen gas (see p. 185). Any precipitate which may form is collected, washed, heated with ammonium sulphide (see p. 186), and the insoluble

portion separated by filtration. The solution is neutralised with acid; if an orange-coloured precipitate forms, it is collected, washed, and tested by the dry tests for antimony (see p. 115). The portion insoluble in ammonium sulphide is well washed and boiled with dilute nitric acid, &c., as directed for the separation of mercury or lead, on page 160.

2. Residue C is to be examined for mercury, lead, and copper.

The filter is perforated, and the precipitate washed with the smallest possible quantity of water into a little conical glass; moderately concentrated nitric acid is added, and a gentle heat applied. If the precipitate entirely dissolves, mercury cannot be present. If anything remain undissolved, pour off the liquid into a small porcelain basin, wash the insoluble matter by decantation (adding the washings to the liquid in the basin), and dissolve it in aqua regia; to this solution apply the copper foil and stannous chloride tests for mercury (see I. (*b*) and (*c*), p. 111).

The nitric acid solution is evaporated on the water-bath almost to dryness. After addition of *pure* sulphuric acid, water is added, and after gentle heating the liquid is decanted off from any white solid matter (lead sulphate). This is transferred to a conical glass, and washed several times by decantation; the liquid and washings being preserved. A strong solution of sodium carbonate is now added, and allowed to stand in contact with the precipitate for several hours, with frequent stirring; a little water is then added and nitric acid until a clear fluid is obtained. The presence of lead in this liquid is confirmed by testing a portion with potassium chromate, and another portion with potassium iodide (see III. (*c*), p. 110). The liquid decanted from the insoluble lead sulphate is concentrated and tested for copper by adding ammonia to one portion and potassium ferrocyanide to another (see III. (*b*) and (*c*), p. 112).

3. Residue D may contain antimony; it is dried, removed from the small filter, which is incinerated, and the ash placed

along with the residue in a porcelain crucible, and fused with potassium cyanide. After cooling, the mass is washed into a porcelain dish, water is added, the washings poured off, and the residue in the dish examined for small metallic globules. If these are found they are heated with a little aqua regia, whereby they are dissolved, fumes being copiously given off. Boil down the solution to drive off nitric acid, dilute with water, and pass sulphuretted hydrogen through a part of the liquid. An orange-coloured precipitate shows antimony. The other part of the liquid is introduced into Marsh's apparatus, and examined for antimony (see pp. 190 and 191).

There now remains only the precipitate by ammonium sulphide (p. 186), which may contain zinc. Acetic acid is added to the liquid in which the precipitate has formed. After a little time the precipitate is collected, washed, dried, and ignited. The residue is dissolved in sulphuric acid, with addition of a little hydrochloric or nitric acid, the liquid freed from excess of acid by evaporation, diluted, and to this solution the usual tests for zinc are applied (see p. 124).

B. *ORGANIC POISONS.*

By the method to be described the suspected matter may be tested for hydrocyanic acid, oxalic acid, and the alkaloids, without introducing any substance capable of interfering with the subsequent tests applied for the detection of the metallic poisons.

I. HYDROCYANIC ACID.—This acid being very volatile, cannot be detected in the stomach, intestines, &c., unless these be subjected to examination soon after death.

The evidence of hydrocyanic acid is generally found in the stomach, intestines, or vomited matter : if not found in any of these it is almost fruitless to search for it in the liver and urine. Its presence has been traced in the brain, spinal marrow, and heart, by the peculiar odour which it possesses.

The first test is the odour emitted by the poison : this resembles bitter almonds. Particular attention should always be paid to this point.

Whether the odour of hydrocyanic acid has been detected or not, it is necessary to proceed with the further examination. The method is based upon the fact that when the matter containing the poison is acidified and distilled, the distillate contains all, or almost all, the hydrocyanic acid. If, however, cyanides, even the non-poisonous ferro- or ferricyanide of potassium, be present, hydrocyanic acid will be found in the distillate, obtained according to this method. A special examination, is, therefore, necessary in order to determine in what form the cyanogen compound originally existed. This examination we shall refer to at a later stage.

The solid matter is cut into shreds, mixed with distilled water, so as to form a thin cream or paste, and, if the reaction be not strongly acid, tartaric acid added in solution ; the whole is then transferred to a retort connected with a condensing apparatus, and gently heated in a calcium chloride bath until 20 c.c. or so have passed over. In this distillate hydrocyanic acid, if present, may be detected by the following tests :—

(*a*) To a part of the distillate a few drops of ferrous sulphate solution and one drop of ferric chloride solution are added, together with a little caustic soda ; a bluish-green precipitate forms, consisting of a mixture of Prussian blue and ferric hydrate ; on adding hydrochloric acid, the ferric hydrate dissolves, leaving a blue precipitate, or, if but very small quantities of the poison are present, a blue-coloured solution, from which a slight precipitate separates after some time.

(*b*) Before applying this test a very small quantity of the distillate is tested with a drop of ferric chloride ; if no red colour is thereby produced—showing the absence of acetates, formates, or meconic acid—a further portion of the distillate is mixed with a few drops of yellow ammonium sulphide, and evaporated at a gentle heat until it has become colourless. Ammonium sulphocyanide is thus formed, which, on acidifying with hydrochloric acid, and adding a drop of ferric chloride, gives a deep red colour (see XV. (*c*), p. 146.)

In applying this test to very small quantities of a fluid, it will be found advisable to place the liquid in a watch glass, over which is inverted a second watch glass, in which a drop of yellow ammonium sulphide has been placed. The lower glass is very gently heated for ten or fifteen minutes, the upper glass is then removed, heated very cautiously, until the liquid which it contains is perfectly colourless, and then one drop of pure hydrochloric acid and a drop of ferric chloride added.

(*c*) Another very good test for small quantities of hydro-cyanic acid consists in immersing in a portion of the distillate a small slip of filtering paper, previously soaked in tincture of guiacum, dried, and moistened with water containing a small quantity of copper sulphate.* The paper is coloured blue if hydrocyanic acid is present.

If the presence of this acid has been proved in the distillate we must institute further experiments to determine whether ferro or ferricyanide of potassium exists in the original substance. For this purpose mix a quantity of the original substance with water, filter, acidulate the filtrate with hydrochloric acid, and to one half add ferric chloride, to the other ferrous sulphate ; the formation of a blue precipitate indicates, in the first case ferro, in the second ferricyanide of potassium. If one or both of these double cyanides have been proved present, there yet remains another problem to be solved, viz. : Did the original matter contain both the non-poisonous cyanide and also the poisonous hydrocyanic

* These papers are prepared by immersing good filtering paper in a solution of 3 parts tincture of guiacum in 100 parts rectified spirits and drying. Just before use, moisten the papers in a solution prepared by dissolving 1 part cupric sulphate in 500 parts of water.

acid or poisonous cyanide ?   Otto recommends the following
process : The mass is mixed with a solution of ferric chloride,
Prussian blue is thus formed if ferrocyanide be present ;
sodium carbonate is added to alkaline reaction, then tartaric
acid, until the mass is slightly acid, and the whole distilled.
If hydrocyanic acid, or a poisonous cyanide were originally
present, the acid will be found in the distillate, but if potassium ferrocyanide only were present, no trace of the acid
will be detected.

II. OXALIC ACID.—In searching for this poison it is
generally necessary to examine not only the stomach,
intestines, &c., but also vomited matter, and, if possible, the
urine and blood ; because the acid being rapidly absorbed
by the system, quickly finds its way from the stomach into
the heart and veins, unless partially ejected in the vomit.

The process for the detection is as follows :—

The substance to be examined is boiled with distilled
water in a large porcelain dish, and the liquid filtered while
hot ; the residue is again boiled and filtered, the two filtrates
united and evaporated to dryness at a gentle heat, the
residue repeatedly treated with small quantities of cold
absolute alcohol, and lastly the alcoholic liquid evaporated
to a small bulk and allowed to crystallise. (If much organic
matter be present in this solution it should be precipitated
by tincture of gall nuts—previously tested for oxalic acid—
and filtered.) The crystals which form are dissolved in water,
and the proper tests for oxalic acid applied to the solution.

The residue remaining after treatment with absolute
alcohol—as first described—must also be tested for oxalic
acid, as it may contain traces of this acid undissolved by the
alcohol : or if an alkaline oxalate were the poison used, it
will be found here.   This residue is treated repeatedly with
small portions of cold water, and the resulting liquid precipitated with calcium chloride ; the precipitate is collected,
washed, and boiled for half an hour or so with a solution of
potassium carbonate, the resulting soluble potassium oxalate

is separated by filtration from the calcium carbonate, lead
acetate is added to the filtrate, and the precipitate which
forms is collected and washed ; it is then suspended in water
through which a current of washed sulphuretted hydrogen
is passed ; after the liquid is thoroughly saturated with the
gas the insoluble lead sulphide is separated by filtration, and
the filtrate evaporated until the acid crystallises out. The
crystals are dissolved in water and the tests for oxalic acid
applied to the solution.

Sometimes antidotes may have been administered in cases
of poisoning by oxalic acid ; those usually employed are
magnesia or a lime salt.   In such instances the detection of
the acid becomes a matter of considerable difficulty.   The
mass remaining after boiling with water is washed with water
in a large dish, the heavier particles being thus separated as
far as possible from the comparatively light organic matter.
The impure oxalate is collected and treated with hydrochloric acid—if magnesia was the antidote employed ; with
moderately concentrated nitric acid—if lime has been used.
The solution is filtered, and saturated with ammonia ; if a
precipitate forms (calcium oxalate) it is digested with acetic
acid to remove traces of calcium phosphate, and then treated
with potassium carbonate, &c., as already described.   If no
precipitate forms on addition of ammonia, potassium carbonate is added, the resulting magnesium carbonate collected
on a filter, and the filtrate treated with lead acetate, &c.

III. ALKALOIDS. — All the more commonly occurring
alkaloids may be detected by the following processes :—

Method A.—Process of Stas, modified by Otto.

The suspected matter is heated to 70° or 75° with twice
its weight of strong alcohol, and acidified with one or two
grammes of tartaric or oxalic acid.

Filter when cold, wash with strong alcohol, and add the
washings to the filtrate. Evaporate the liquid, either under

a bell jar over strong sulphuric acid, or in a retort through which a current of air is drawn by means of the filter pump.

If fatty or insoluble matter separates during this process, remove it by filtration, and continue to evaporate the liquid nearly to dryness. Digest the residue with *cold absolute* alcohol, filter, wash the residue with alcohol, and evaporate the filtrate to dryness in vacuo, or spontaneously in the air. Dissolve the residue in the smallest possible quantity of water, and shake this solution with ether repeatedly, so long as the ether becomes coloured and yields a residue on evaporation ; add now acid sodium carbonate so long as effervescence ensues, again add 4 or 5 volumes of ether, shake up well, and allow to stand. Transfer a small quantity of the ethereal liquid to a watch glass and allow it to evaporate spontaneously. If oily drops are left, which emit a pungent odour when heated, volatile bases are probably present ; if a solid residue, or a turbid fluid with solid particles suspended in it remain, a non-volatile base must be looked for.

If it is necessary to search for a volatile base, add to one half of the mixed ethereal and aqueous solutions 1 or 2 cb. cm. of strong caustic potash solution, and shake well ; the base is thus liberated and is dissolved by repeated treatment with ether, which must be continued until the lower portion leaves no residue on evaporation. Dilute sulphuric acid is now added to the ethereal solution to *decidedly* acid reaction ; shake well, and after decanting the supernatant ether, again treat the aqueous fluid with dilute acid.

The aqueous solution now contains the bases, ammonium, nicotine, picoline, and the greater part of the conine, if present, as sulphates. By addition of strong caustic potash in excess, and subsequent treatment with ether, these bases are again liberated and dissolved, and on evaporating the ethereal liquid (best *in vacuo* over sulphuric acid) they are obtained in the residue. Ammonia is removed by this evaporation. *Call this Residue A.*

From the second half (or from the whole if no volatile

bases are present) of the mixture of aqueous and ethereal
solutions, the ether is decanted and allowed to evaporate.
The residue is dissolved in a small quantity of water
acidulated with sulphuric acid, the solution shaken repeatedly
with ether, and the ether removed (animal or fatty matter is
thus got rid of). Sodium carbonate is now added in excess
to the aqueous solution, and the liquid again shaken with
ether ; the resulting ethereal liquid, on spontaneous evaporation, yields the non-volatile bases nearly pure. *Call this
Residue B.*

Method B.   Process of Uslar and Erdmann.

This process is more suited for the detection of non-
volatile alkaloids. The suspected matter, mixed with water
to the consistence of a thin paste, and acidified with dilute
hydrochloric acid, is heated for several hours to 70° or 80°
then strained through moistened linen, the residue washed
with hot acidulated water, and the washings added to the
filtrate. The liquid is now saturated with ammonia and
evaporated to dryness. By the addition of a little pure sand
the residue is easily reduced to powder. Boil the powder
repeatedly with amyl alcohol, and filter while hot. To the
alcoholic solution, transferred to a cylinder, add ten volumes
of almost boiling water, acidify with hydrochloric acid, and
shake the mixture vigorously; the alkaloids—as hydrochlo-
rates—pass into the aqueous solution while the amyl alcohol
retains the fatty matter. The latter is removed by a pipette,
and the acid solution again shaken with amyl alcohol to
remove the last traces of colouring matter. By concentrating,
treating with ammonia, and repeated shaking with hot amyl
alcohol, the bases will be obtained in the alcoholic solution
nearly pure.

If any further purification is required, the residue is
redissolved in hydrochloric acid, after removal of the amyl
alcohol by evaporation on the water bath; the liquid is

shaken with alcohol, the latter drained off, and ammonia added, &c., as described. The residue, after removal of the alcohol, contains the pure alkaloids—a small portion of it should yield no brown colour with a drop of strong sulphuric acid.

We now proceed to the further testing of the several residues.

Residue A (p. 201) contains the volatile alkaloids.

The presence of conine or nicotine in this residue will be made manifest by its odour, that of the former being very strong and penetrating, and that of the latter ethereal and tobacco-like. The odour of picoline is also very characteristic ; when inhaled through the nostrils the vapour of this substance produces a bitter taste in the mouth and back of the throat. On the addition of water to this residue, nicotine, if present, dissolves, while picoline or conine remains insoluble. If, therefore, oily drops form on the surface of the water, one or other of the latter alkaloids is present. If no such drops form, nicotine must be sought for.

To distinguish between conine and picoline, add to small quantities of the oily liquid on watch glasses—

1. Strong nitric acid. A blood-red colour shows the presence of CONINE.

2. Solution of albumen. It is coagulated by CONINE.

3. Add a few drops of hydrochloric acid, and evaporate gently, then add platinum tetrachloride. If a yellowish crystalline precipitate forms, PICOLINE is present ; but if this precipitate is produced only on addition of alcohol, CONINE is present.

To distinguish between nicotine and picoline ; mix small quantities of the watery solution in watch glasses with—

1. Hydrochloric acid and platinum tetrachloride. A precipitate indicates NICOTINE ; if, however, the precipitate only forms upon evaporation, this shows the presence of PICOLINE.

2. Tincture of iodine added to solution of NICOTINE gives

a yellowish turbidity, which disappears after a little time ; but on further addition of the reagent a copious kermes-coloured precipitate separates, which also disappears upon standing.

Residue B (p. 202) contains the non-volatile alkaloids. Generally the examination is limited to a search for one particular alkaloid. Of these, which may be here detected, we shall notice morphine, narcotine, brucine, strychnine, veratrine, colchicine and aconitine. If we are searching specially say for strychnine, then the tests for this alkaloid are to be at once applied to a portion of the crystalline residue. If, however, we possess no clue to the nature of the alkaloid present, we must place several fragments of the residue in small clean porcelain dishes, and apply the following tests :—

1. 4 or 5 drops of *pure strong* sulphuric acid.

A yellow colour changing to red or violet indicates VERATRINE or ACONITINE.

A bluish-violet or yellow, changing on heating to cherry red, indicates NARCOTINE.

A rose colour changing to yellow indicates BRUCINE.

A yellow-brown colour indicates COLCHICINE.

Add now to the fluid obtained as above—

2. 10 or 20 drops strong sulphuric acid containing nitric acid,* after 20 to 30 minutes if there ensues

A violet colouration, it indicates MORPHINE.

STRYCHNINE is not coloured by any of these reagents.

From these tests an indication as to which alkaloid is present will have been obtained. Confirmatory tests must now be applied. The following are very good :—

I. VERATRINE ($C_{32}H_{52}N_2O_8$).—

Strong sulphuric acid at first causes this alkaloid to form into small lumps ; these, however, speedily dis-

---

* Three drops of strong nitric acid are mixed with 100 c.c. of water, and ten drops of this liquid added to 40 c.c. of pure strong sulphuric acid.

solve to a light yellow fluid, changing to reddish yellow, and eventually to blood red.

Strong hydrochloric acid forms a colourless solution, which becomes intensely red by long-continued boiling. This reaction is very delicate.

II. ACONITINE $(C_{30}H_{47}NO_7)$.—

Sulphuric acid produces a yellow colour changing to violet.

Nitric acid dissolves it without colour.

III. NARCOTINE $(C_{22}H_{23}NO_7)$.—

The reaction already given is the most delicate for this alkaloid. The changes of colour are marked, at first a bluish violet (or in less pure specimens a yellowish colour) is produced. On gently warming, the colour changes to orange-red, bluish stripes then appear ; if the heating is now discontinued the liquid becomes cherry-red as it cools, but by continuing the application of heat a deep violet-red is formed.

IV. BRUCINE $(C_{23}H_{26}N_2O_4)$.—

Strong nitric acid yields a very red fluid, becoming yellow when heated, and changing to intense violet on addition of stannous chloride or ammonium sulphide.

Strong sulphuric acid (4–5 drops) produces a faint rose-colour, changing to yellow ; 10–20 drops of sulphuric acid mixed with nitric acid (see note, p. 204) being now added, a transient red colour, changing again to yellow, is produced.

V. COLCHICINE $(C_{17}H_{19}NO_5)$.—

Strong nitric acid gives a deep violet or blue colour, changing to olive-green and yellow.

Tincture of iodine throws down a kermes-coloured precipitate.

Tannic acid gives a white precipitate, soluble in alcohol and acetic acid.

VI. Morphine ($C_{17}H_{19}NO_3$).—

The test already mentioned, viz., the formation of a violet colour upon addition of sulphuric acid followed by a mixture of the same acid with nitric acid, is characteristic.

If a few small pieces of manganese dioxide are added to the liquid obtained as above, it changes to mahogany colour.

If solid morphine or a morphine salt is moistened with a solution of 1 part of iodic acid in 15 parts of water, and a solution of 1 part of starch in 400 of water is added, a blue colour is produced. If now a layer of very dilute ammonia is poured on this blue liquid, two coloured rings are formed at the point of contact, the upper blue, the lower brown. This reaction serves to detect $\frac{1}{20,000}$th part of morphine (*A. Dupré*).

A neutral concentrated solution of a ferric salt gives a deep blue colour with salts of morphine.

VII. Strychnine ($C_{21}H_{22}N_2O_2$).—

Strychnine and its salts have an *intensely bitter* taste.

A colourless solution of this alkaloid in *pure* sulphuric acid shows a splendid bluish colouration, changing to wine-red and then to reddish-yellow, on addition of the following oxidising agents :—

　　1. A very small quantity of solid manganese dioxide.
　　2. A crystal of potassium chromate.
　　3. A crystal of potassium permanganate.

The action of the two last reagents is very marked. The small fragment of the oxidiser used should be pushed about the dish with a glass rod.

If morphine be present it interferes with these reactions, and must therefore be removed. To do this, add potassium chromate to the concentrated *aqueous* solution, and filter (the morphine is found in the filtrate) ; wash the precipitate slightly, dry it, and mix it in a watch glass with a few drops of strong pure sulphuric acid ; the blue violet colour is immediately produced.

The following are a few special methods for the testing of certain alkaloids.

I. QUININE ($C_{20}H_{24}N_2O_2$).

The acid salts of this alkaloid exhibit *fluorescence* in a remarkable degree.

Tests. Chlorine water with subsequent addition of ammonia gives an intense emerald-green colour. If the addition of ammonia be preceded by that of a few drops of a solution of potassium ferrocyanide, a deep red tint is produced, soon turning to a dirty brown.

Bromine vapour may be advantageously substituted for chlorine in the above reactions (*Flückiger*).

Medicinal *sulphate of quinine* often contains as impurities small quantities of the sulphates of cinchonine and quinidine. To detect these, the method of *Liebig* as modified by *Stoddart* is employed : ·6 grm. of the suspected salt is dissolved in about 10 drops of pure dilute sulphuric acid, with the addition of 50 or 60 drops of distilled water in a small stoppered bottle ; 150 drops of pure ether, mixed with 3 drops of alcohol and 40 drops of a solution of caustic soda (1 to 12 of water), are now added. After brisk agitation the bottle is set aside for 12 hours, when the smallest traces of quinidine or cinchonine, if present, may be seen separating out at the line of separation of the two liquids in the bottle.

DETECTION OF STRYCHNINE IN BEER (Graham and Hofmann). One gallon of the beer is shaken with 4 ounces of animal charcoal, and after standing 12 hours the latter is collected and washed once or twice with cold water, then boiled with alcohol, the alcoholic solution evaporated to dryness at a gentle heat, the residue shaken with a few drops of caustic potash and ether, the ethereal solution evaporated, and the residue tested for strychnine by the usual methods.

DETECTION OF ADMIXTURE OF BRUCINE WITH STRYCH- NINE.—The presence of brucine in samples of strychnine may be detected by the intense red colour imparted to the nitric acid solution of the suspected salt when the former alkaloid is present.

DETECTION OF OPIUM.—As opium contains meconic acid $(C_7H_4O_7)$ in addition to morphine, it is necessary to obtain the tests for this body before deciding whether a given sample actually does or does not contain opium.

Meconic Acid.

1. Ferric chloride produces a deep red colouration, un- altered by boiling or treatment with mercuric chloride.

2. Lead acetate precipitates white lead meconate, in- soluble in acetic acid.

To prove the presence of opium in organic tissues, &c.: evaporate the suspected matter to dryness on the water- bath, and digest with water containing a little acetic acid for some time ; filter, and add lead acetate to the filtrate, warm and again filter after cooling.   From the filtrate, excess of lead is removed by means of sulphuretted hydrogen, the clear liquid evaporated to dryness, a little potassium carbo- nate added, and the mixture shaken with an ethereal solution of acetic ether ; this solution will now contain the morphine, which may be tested for according to VI. p. 206.

The precipitate produced by lead acetate is suspended in water, through which a current of sulphuretted hydrogen is passed, the resulting lead sulphide removed by filtration, excess of sulphuretted hydrogen driven off by heat (the tem- perature must not exceed 70° C.), and the concentrated liquid tested for meconic acid.

IDENTIFICATION OF BLOOD STAINS.

If the stains occur in undyed cloth, linen, or cotton, the stained piece is cut out, and suspended in pure water until the stain is almost entirely removed. To the solution thus obtained various tests are applied.

1. On heating, the red or reddish-brown colour disappears, while small particles of coagulated albumen are seen floating about the liquid ; these disappear on addition of caustic soda.

2. A solution of tannin gives a white precipitate.

3. A portion evaporated in a porcelain dish leaves a brown lustrous coating which, on treatment with chlorine water and subsequent evaporation, becomes colourless. If a few drops of water and a very little potassium sulphocyanide are now added a reddish colour is produced, owing to the iron in the blood.

4. A small quantity of tincture of guiacum mixed with an ethereal solution of hydrogen peroxide, gives a *blue colour* (this colour is produced with other substances than blood).

5. Another part of the stained material is boiled with glacial acetic acid, a little sodium chloride being added during the ebullition ; a few drops are then evaporated to dryness at from 40° to 60°, and the residue examined under the microscope for hœmin crystals. This compound forms rhombic crystals, or sometimes six-sided plates, dark blue by reflected, dirty brown by transmitted light. If the stain occurs in soil or on iron, the dried soil, or the part scraped off the iron, must be heated with weak potash ley, and chlorine water added to the filtered solution. A coagulum is thus formed which yields hœmin crystals on treatment with salt and glacial acetic acid. Some observers say that the hœmin crystals from the greater number of warm-blooded animals are identical ; while, according to others, those obtained from human blood differ essentially from all others.

P

Blood gives very characteristic absorption bands with the spectroscope. The process, with a description of the instruments, &c., to be employed, and of the results obtained will be found fully described in a series of papers by Mr. H. C. Sorby, in the *Chem. News*, vol. xi., 1865.

SECTION V.

EXAMINATION OF URINE AND URINARY CALCULI.

THE number of substances contained in healthy urine is very great ; for the purpose of qualitative examination they may be roughly classed as follows :—

1. Water. 2. Inorganic substances—Potassium, sodium, calcium, magnesium, and iron, existing as chlorides, phosphates, sulphates, nitrates, carbonates, or silicates. Ammonia also occurs in small quantities. 3. Organic substances.—*a*. Nitrogenised ; urea, uric and hippuric acids ; creatinine and xanthine ; —*b*. Non-nitrogenised ; lactic, succinic, oxalic, formic, and phenylic acids, and sugar (?). 4. Colouring and extractive matter, and more or less mucus in suspension.

FIG. 57.

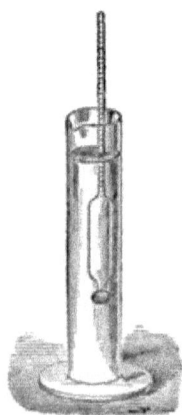

In addition to these small quantities of many other substances are found in healthy urine, while in cases of disease there may be present—blood, albumen, fibrine, fats, cystine, glucose, &c. The reaction of the urine should be first ascertained by test-papers ; healthy urine has generally an acid reaction, but immediately after meals it may be neutral or even slightly alkaline. The specific gravity of healthy urine varies from 1·015 to

1·020. This is determined by means of the urinometer (see fig 57), which consists of a glass bulb, containing mercury, to sink it in the liquid, furnished with a stem, graduated from 1,000 to 1,060. A sample of the urine, which should represent the average quality of the entire amount voided during twenty-four hours, is poured into a cylindrical glass, the urinometer floated in the liquid, and the number of degrees at the point where the surface of the urine is in contact with the stem, read off, and noted down. The urinometer is now depressed in the urine, and after coming to rest another reading is taken. A high specific gravity generally points to the presence of sugar, a low specific gravity is more frequently due to the ingestion of large quantities of liquid, but it is also caused by several diseases.

2. The odour, transparency, and colour of the urine must also be noted. If the colour is saffron-yellow, or inclining to brown, probably biliary matter is present ; in this case a yellow scum forms on the surface on agitating the sample. If the urine is turbid this is probably due to the presence of fatty or chylous matter. A smoky-red colour is diagnostic of blood.

3. A quantity of the urine is allowed to stand for some time ; if a sediment forms it is collected on a filter and reserved for examination.

4. The clear filtrate must now be tested for :

a. Albumen.—Heat the upper layer of a small quantity of the sample in a test-tube to the boiling point, the lower layer remaining cool ; by now holding the tube against the light the slightest turbidity in the upper layer is easily detected. Allow a few drops of dilute nitric acid to flow down the tube into the urine. If the cloudiness is permanent it is due to albumen which has been coagulated by the heat ; if the cloudiness disappears, it is due to the presence of earthly phosphates ; no albumen is present. If the preliminary examination has shown that the urine has an alkaline reaction it must be rendered slightly acid by the addition of a few drops of *acetic acid* before testing for albumen.

In the event of albumen being detected, the whole of the urine, filtered from sedimentary matter, must be heated with the addition of a little acetic acid, filtered, and the filtrate tested for:

b. Urea and Uric Acid.—A portion of the sample is concentrated on the water-bath until it is of the consistence of a syrup. After the liquid is cold, nitric acid (free from nitrous acid) is gradually added; the formation of delicate rhomboidal crystals of nitrate of urea, which are best seen under the microscope, tells us that urea is present.

Another portion of the sample, placed in a beaker glass, is acidulated with hydrochloric or acetic acid, and allowed to stand for 30 or 40 hours. If reddish-brown crystals form on the sides or bottom of the glass they are removed and heated with nitric acid; reddish-coloured fumes are evolved, while the nitric acid solution, when evaporated to dryness, gives, on the addition of strong ammonia water, a magnificent purple colouration. These reactions are indicative of uric acid. (The colouration is caused by the formation of murexide, $C_8N_6H_8O_6$.)

c. Hippuric Acid ($C_9H_9NO_3$).—A quantity of the urine is evaporated to about the third of its volume on the water-bath, and allowed to cool. If a crystalline precipitate forms it may contain calcium hippurate (in addition to other substances). This acid occurs in very small quantities in urine; as its clinical import is unknown, further details of its detection are unnecessary.

Chlorides, phosphates, and sulphates are also present in healthy urine. Their presence is recognised by the usual tests.

The above-mentioned substances, with the exception of albumen, are normal constituents of urine; we now proceed to test for certain substances found in abnormal urine.

a. AMMONIUM CARBONATE.—A quantity of the clear

filtered urine is treated with hydrochloric acid ; effervescence ensues. Another portion is boiled with caustic potash, when ammonia is disengaged.

b. SUGAR.—Prepare a little of Fehling's copper solution by dissolving 3 grams of pure dry copper sulphate in 20 c.c. of water, and adding gradually a solution of 1·5 grams of crystallised sodium-potassium tartrate in 6 c.c. of caustic soda solution (sp. gr. 1·14). This solution must be kept in a blue stoppered bottle.

Heat a few cubic centimetres of this solution to boiling in a test tube, and gradually add the sample of urine drop by drop ; if a red precipitate falls down (cuprous oxide, Cu_2O) sugar is present. Before using this solution it must be heated to boiling and carefully examined ; if any precipitate forms the solution is unfit for use—a new one must be prepared. The amount of urine added should not exceed the bulk of Fehling's solution used.

c. SULPHURETTED HYDROGEN. — This gas very rarely occurs in fresh urine ; its presence is recognised by its smell and its blackening action on slips of paper moistened with lead acetate.

The sedimentary matter which may have formed when the urine was allowed to stand at rest must now be examined. It may contain (*a*) uric acid, generally appearing in a granular precipitate varying in colour from light yellow to orange yellow and brown. (See reaction with nitric acid and ammonia, p. 212.)

d. URATES.—The sediment caused by urates (chiefly of sodium, potassium, calcium, and ammonium) varies in appearance from acicular crystals to irregular clot-like masses, hardly distinguishable from mucus or blood.

The sodium and potassium urates are soluble, the calcium and ammonium salts almost insoluble, in water. A drop of hydrochloric acid added to a sediment composed of urates sets free uric acid, which may be recognised under the microscope by its crystalline form, and by the formation of murexide when treated with nitric acid and ammonia.

e. CALCIUM OXALATE.—This forms a sediment, in which 'dumb-bell' shaped crystals are occasionally visible. Calcium oxalate is insoluble in acetic acid. After gentle ignition on platinum foil calcium carbonate remains, which effervesces on addition of an acid. Dissolve a little of the supposed calcium oxalate in nitric acid, and add mercurous nitrate, mercurous oxalate will be precipitated ; this is removed by filtration, ammonia added to the filtrate to precipitate excess of mercury (existing as nitrate), and after another filtration oxalic acid is added, when a white precipitate of calcium oxalate is obtained.

f. MAGNESIUM - AMMONIUM PHOSPHATE, OR CALCIUM PHOSPHATE.—The former double phosphate occurs in large transparent rhombic prisms, the latter generally in an amorphous powder, but sometimes in small penniform, or acicular crystals. Dissolve the sedimentary matter in a little hydrochloric acid, add sodium acetate in excess, calcium oxalate, if present, is thus precipitated ; filter, and to the filtrate add oxalic acid, which precipitates the calcium which originally existed as phosphate, again filter, and add to the filtrate ammonia in excess, when magnesium phosphate will be thrown down.

EXAMINATION OF URINARY CALCULI.

These calculi may be divided into groups according as they comport themselves when heated on platinum foil. Thus we have the following divisions :—

I. *Calculi which leave no fixed residue.*—These may be composed of uric acid, ammonium urate, or hippurate, xanthine, cystine, cholesterine, colouring matters of the bile, fibrine, or albumen.

II. *Calculi which leave a fixed residue.*—This group is subdivided into

a. Those which are partially combustible.—Sodium or

calcium urates, and calcium or magnesium phosphates, oxalates, or carbonates.

b. Those which are entirely fixed.—These must consist of mineral matter entirely.

Group I.—*Calculi which leave no fixed residue.*

A. A portion of the substance is heated with concentrated nitric acid, the resulting solution evaporated to dryness on the water-bath, and the residue moistened with one drop of strong ammonia water.

a. It is coloured purple-red.—Uric acid, or ammonium urate. Another small portion of the calculus is gently heated with caustic potash.

    a. Ammonia is evolved.—AMMONIUM URATE.

    β. No ammonia is evolved.—URIC ACID.

b. It is not coloured purple-red.

    a. The nitric acid solution become *brown* during evaporation. The residue is soluble in ammonia, microscopic hexagonal tables being deposited from this solution.—CYSTINE.

    Cystine when heated evolves a peculiar empyreumatic odour. The potash extract, when decomposed by a solution of lead oxide (PbO) in caustic potash, and heated, gives a precipitate of lead sulphide if cystine were present (*Liebig*). (Mucus, albumen, and other sulphur-containing compounds would, however, give a similar reaction.)

    β. The nitric acid solutions become yellow during evaporation. The residue is amorphous, and insoluble in hydrochloric and oxalic acids.—XANTHINE.

B. A portion of the calculus is heated on platinum foil, It burns with a clear white flame. — CHOLESTERINE or FIBRINE.

    a. If the calculus possesses a crystalline structure, dissolves in boiling alcohol, from which solution small shining spangles are precipitated, and is insoluble in, and not saponified by caustic potash, it consists essentially of CHOLESTERINE.

b. If during combustion of the calculus the smell of burning horn is perceptible ; if a portion of the calculus dissolves in caustic potash, and is precipitated from this solution by acetic acid, but is soluble in excess of this reagent, it consists chiefly of FIBRINE.

C. The calculus is of a brown colour, it is friable, and when burning the smell of calcined animal matter is perceptible. BILIARY ACIDS, or COLOURING MATTER.

a. It is scarcely soluble in alcohol or water, but is dissolved (with a brown colour) by caustic potash. If, on adding to this solution an equal volume of alcohol, and then a little *fuming* nitric acid, the colour changes to green, then to blue, violet, and red, after some time again turning yellow, we conclude that BILIARY COLOURING MATTER is present.

b. It is soluble in alcohol. This solution has a bitter taste, and when mixed with one drop of a sugar solution (one part sugar to four of water) and pure strong sulphuric acid added drop by drop (cooling whenever the mixture gets hot), it shows a yellowish tint, changing to pale cherry red, and eventually to an intense violet.—BILIARY ACIDS.

Group II.—*Calculi which leave a fixed residue.*

A. The calculus gives with nitric acid and ammonia the reaction characteristic of uric acid (see *b.* p. 212).

To determine whether sodium, potassium, calcium, or magnesium be present in combination with this uric acid, notice the behaviour of the calculus when heated.

a. It melts.

α. It gives a deep yellow colour to the Bunsen flame. —SODIUM URATE.

β. It gives a violet colouration to the flame.—POTASSIUM URATE.

b. It does not melt.

α. After calcination calcium carbonate remains (recognisable by its effervescence with an acid, &c.)—CALCIUM URATE.

β. After calcination the residue dissolves with *slight* effervescence in dilute sulphuric acid. This solution neutralised with ammonia gives on addition of sodium phosphate a white precipitate.—MAGNESIUM URATE.

B. The calculus does not give the characteristic uric acid reaction. Calcine a portion of the calculus and examine the residue.

a. The residue melts on heating.—It probably consists of CALCIUM or MAGNESIUM-AMMONIUM PHOSPHATES.

α. The calculus does not effervesce with acids either before or after calcination. From a hydrochloric acid solution ammonia and ammonium oxalate both throw down a white precipitate.—CALCIUM PHOSPHATE.

β. During the calcination an ammoniacal odour is perceptible. The residue dissolves without effervescence in acetic acid ; from this solution ammonia throws down a white crystalline precipitate.—MAGNESIUM-AMMONIUM PHOSPHATE.

These two phosphates may also be recognised by the reactions given under the head of examination of sedimentary deposits in urine (p. 213.)

b. The residue does not melt on heating. The calculus may contain CALCIUM PHOSPHATE, OXALATE, or CARBONATE.

    a. The residue is white, and does not turn red litmus blue. Dissolved in a little hydrochloric acid and sodium acetate, and *one drop* of ferric chloride added, a dirty white precipitate (ferric phosphate) is formed. To the filtrate from this precipitate ammonium sulphide is added, and excess of iron thus removed ; after filtration the addition of oxalic acid causes a white precipitate.—CALCIUM PHOSPHATE.

    β. The calculus dissolves in the mineral acids without effervescence, but after calcination these acids dissolve the residue with evolution of carbon dioxide. The reactions detailed on p. 127 for calcium oxalate are also given.—CALCIUM OXALATE.

    γ. Mineral acids dissolve the calculus with effervescence ; from this solution neutralised with ammonia, ammonium oxalate throws down a white precipitate.—CALCIUM CARBONATE.

If doubt should arise as to the presence of any of these substances in a calculus, special confirmatory reactions must be resorted to.

APPENDICES.

A.—*Apparatus List.*

Test-tubes. Test-tube stand. Test-tube holder. Test-tube brush. Steel spatula. A small porcelain, horn, or platinum spatula is also very useful. Small Wedgwood mortar and pestle.

Three or four porcelain basins varying from 5 to 10 centimetres in diameter.

Half a dozen glass rods in length about 20 centimetres, rounded at the ends by slight fusion in the blow-pipe flame.

Half a dozen flasks; these should include both 'Florence flasks,' which are round bottomed and have no lip, and also flat-bottomed lipped flasks. The sizes should vary from 100 to 500 c.c. capacity.

One or two clamps fitted with bosses.

Metal stand with three rings.

Iron tripod stand.

One or two small Bunsen lamps about 4 in. high, and the tubes about $\frac{3}{8}$ in. diameter.

Wire gauze cut in squares.

Pneumatic trough; the round stoneware troughs with perforated moveable beehives are very serviceable.

Small trays for holding gas bottles; these may be made of thin sheet zinc; the lids of tin canisters will serve very well.

Gas bottles; these bottles should be about 20 or 25 cm. high, and 5 to 7 cm. in diameter, with wide necks.

Deflagrating spoons.

Funnel-tubes; these are sometimes called 'thistle-headed tubes.'

A separating funnel, with stop-cock, to hold about 100 c.c.

Filtering-paper.—Imitation Swedish will do very well for qualitative analysis. It will be found convenient to have the paper cut into circular pieces, having diameters varying from 2 to 8 centimetres.

Caoutchouc tubing and corks; the red tubing is preferable to grey, as it does not split so easily.

Glass-plates, also round pasteboard discs, for covering gas bottles &c.

Bell jar ; this should hold about 750 cb. cm.

Retorts ; one or two of capacities varying from 50 to 200 cb. cm.

Files, round and triangular.

Stout soda-water bottle.

Large test-tubes, 20 cm. long, $2\frac{1}{2}$ cm. diameter.

A few small beakers. Nos. 2 to 5 are the most convenient.

Glass tubing of various sizes.

A glass cylinder, about 40 cm. in height and 8 cm. in diameter.

A Wolff's two-necked bottle.

Calcium chloride tubes.

Pipettes, to hold 5, 10 and 20 c.c.

Crucibles of porcelain, crucible tongs and triangles, also a few Hessian crucibles.

One or two candles.

Copper wire and a pair of pliers.

Platinum wire moderately thick, about 40 cm. in length.

This list will be found to comprise the greater part of the necessary apparatus.

B.—*Reagent List.*

The following chemical reagents are required in the first part of the course.

Potassium chlorate.

Manganese dioxide.

Silver nitrate.

Flowers of sulphur.

Charcoal.

Phosphorus.

Lime water ; this is prepared by agitating recently slaked lime with cold water, allowing the undissolved portion of the lime to subside and drawing off the clear liquid ; it must be kept in a well stoppered bottle.

Granulated zinc, prepared by dropping molten zinc from a height into cold water.

Strong sulphuric acid.

Nitric acid.

Hydrochloric acid.

Sodium chloride (common salt).

Powdered metallic antimony.

Turpentine.

Copper foil.

Ferrous sulphate.

Ammonium nitrate.
Ammonium chloride.
Caustic lime.
Calcium carbonate (marble).
Oxalic acid.
Ammonia water.
Caustic potash solution.
Alcohol (methylated spirit).
Sodium acetate and caustic soda.
Potassium bromide.
Potassium iodide.
Ether.

Chlorine water; prepared by passing chlorine gas into cold water until the latter smells strongly of the gas.

Starch paste; prepared by boiling a small quantity of starch in water, and allowing the liquid to cool.

Amorphous or red phosphorus.
Iodine.
Calcium fluoride (fluorspar).
Mercury.
Barium chloride solution.
Platinum tetrachloride solution. (See p. 223, No. 24.)
Benzene.
Calcium chloride.
Cane sugar.
Iron pyrites.
Potassium dichromate.
Ink—writing and printers'.
Carbon disulphide.
Sodium carbonate.
Sand, bees-wax, tapers, Turkey-red cloth and madder-dyed cloth.
Asbestos.

The following reagents are required in the course of qualitative analysis.

I. *Ordinary Reagents.*

1. **Hydrochloric acid**: this is prepared by adding three parts of water to one part of the strong acid.

2. **Sulphuric acid**: one part of the strong acid is diluted with four parts of water, the mixture allowed to cool, and the clear liquid decanted from any white sediment of lead sulphate.

3. **Nitric acid**: one part of the strong acid diluted with three parts of water.

4. **Aqua regia,** prepared by adding four parts of strong hydrochloric acid to one part of strong nitric acid.

5. **Ammonia**: one part of strong liquor ammoniæ is diluted with three parts of water.

6. **Sulphuretted hydrogen.** This reagent is preferably used in the gaseous state, and must be prepared as required ; it is advantageous, however, to saturate a bottle filled with water with the gas (passing it in until on shaking the bottle the smell of sulphuretted hydrogen is very apparent), and keep this closely stoppered and as much as possible out of the light.

7. **Ammonium sulphide.** Prepared by passing sulphuretted hydrogen through strong ammonia water until it ceases to give a precipitate with magnesium sulphate ; a little sulphur may then be added, and the liquid diluted with one part of water.

8. **Ammonium oxalate**: one part by weight, is dissolved in twenty-four parts by measure, of water, and the solution filtered.

9. **Ammonium chloride**: one part by weight in eight measured parts of water.

10. **Ammonium carbonate**: one part in eight measured parts of water.

11. **Potassium ferrocyanide**: one part by weight in twelve measured parts of water.

12. **Sodium phosphate**: dissolve one part by weight, in ten parts by measure, of water.

13. **Barium chloride**: the same proportions as in the last.

14. **Magnesium sulphate**: one part in eight of water.

15. **Caustic potash,** or **soda**: one part in eight of water.

II. *Reagents less frequently used.—Special Reagents.*

1. **Tartaric acid**: this reagent is generally used in a saturated aqueous solution.

2. **Acetic acid**: glacial acetic acid may be diluted with about two parts of water.

3. **Sulphur dioxide** must be prepared (by heating copper clippings with strong sulphuric acid) when required.

4. **Oxalic acid** (generally used as a solid reagent) : the solution is prepared by dissolving one part by weight, in twenty measured parts of water.

5. **Ammonium nitrate**: used in the solid state.

6. **Ammonium molybdate** in nitric acid : one part of ammonium

molybdate is dissolved by heating in three parts of ammonia, and the solution poured into twenty parts of dilute nitric acid (one part acid to one of water).

7. **Potassium ferricyanide** : a solution of one part by weight of this salt in twelve parts by measure of water must be made when required for use. It decomposes slightly on keeping.

8. **Potassium nitrate** : prepared by saturating caustic potash (sp. gr. 1·27) with nitrous fumes—evolved by the action of eight parts of nitric acid (sp. gr. 1·4) on two parts of starch—filtering, evaporating to dryness and dissolving the dry salt in two parts of water.

9. Potassium sulphocyanide : one part in ten parts of water.

10. Potassium chromate : one part in eight parts of water.

11. **Potassium cyanide** : one part in eight parts of water.

12. **Sodium acetate** : dissolve one part in four parts of water.

13. **Borax** : employed in the solid state in flame reactions.

14. **Microcosmic salt** (sodium-ammonium-hydrogen phosphate) is also used as a flux.

15. **Calcium sulphate** : shake up the salt with cold water for a considerable time, and filter.

16. **Barium carbonate**, is mixed with water to the consistence of cream ; shake well before using.

17. **Ferric chloride** : prepared by dissolving recently precipitated ferric hydrate in pure hydrochloric acid, keeping the salt slightly in excess. The solution is diluted with an equal volume of water, and filtered.

18. **Lead acetate** : one part is dissolved in ten measured parts of water.

19. **Lead dioxide** : used in the solid state.

20. **Mercuric chloride** : one part in sixteen parts of water.

21. **Ammonio-copper sulphate**, is prepared by adding ammonia drop by drop to a tolerably strong solution of copper sulphate until the precipitate which forms is nearly all dissolved. Filter, and preserve the clear solution for use.

22. Ammonio-silver **nitrate** : prepared by adding ammonia to silver nitrate in the same way as directed above.

23. **Cobalt nitrate** : one part is dissolved in ten parts of water.

24. Platinum **tetrachloride** : spongy platinum, or platinum foil in little pieces is boiled in nitric acid, washed, and dissolved by heating with aqua regia ; the solution is evaporated to dryness on the water-bath, hydrochloric acid added, and the liquid again evaporated to dryness. The residue is dissolved in about ten parts of water and the solution filtered.

25. Stannous chloride. Dissolve the salt by boiling in strong hydrochloric acid, dilute with four parts of water, and keep the liquid in a stoppered bottle containing a little granulated tin.

26. Sodium hypochlorite (chloride of soda). Dissolve one part by weight of bleaching powder in one and one-sixth parts, by measure, of water, and shake this with a solution of two parts sodium carbonate (crystals) in about one-fourth part by measure of water; filter, and preserve the clear liquid for use.

27. Nessler's reagent is made by dissolving 5 grams of potassium iodide in 18 c.c. of water, adding a cold concentrated solution of mercuric chloride until the precipitate of mercuric iodide which forms just ceases to be dissolved on stirring. Now add an aqueous solution of caustic potash prepared by the solution of 15 grams of caustic potash in 30 c.c. of water. The liquid is diluted to about 100 c.c., and preserved in a stoppered bottle for use.

28. Zinc and copper : the former in small rods, the latter in thin slips, are required for certain tests.

29. Litmus papers, are either blue or red; the former are prepared by boiling one part of commercial litmus in six parts of water; filtering the solution, dividing the filtrate into two parts, neutralising the free alkali in one part by stirring with a glass rod dipped in dilute sulphuric acid until the liquid appears faintly red, and then adding the other part. Slips of unglazed paper are drawn through this solution and then hung on a thread, out of the way of acid fumes, until dry; these slips are kept in a stoppered bottle. Red litmus papers are prepared by soaking unglazed paper in a solution of litmus which has been reddened by stirring with a rod dipped in sulphuric acid.

30. Turmeric papers are prepared by soaking slips of paper in a solution made by heating one part of powdered turmeric root in six parts of dilute spirit of wine. The papers when dried have a pure yellow tint.

C.—*Preparation of Pure Reagents.*

1. Hydrochloric acid may be purified by dilution and redistillation. In order to oxidise sulphurous acid the distillate is treated with chlorine water until the liquid turns starch paper moistened with potassium iodide blue. Commercial tin salt (stannous chloride) is now added in the proportion of five grams to every kilo. of the acid; after standing for a day or two all the arsenic is separated. The acid is finally distilled after addition of a little sodium chloride.

2. Nitric acid, is freed from nitrogen peroxide by passing a current of carbon dioxide through the liquid heated nearly to the boiling point in a retort. A little potassium nitrate is then added, and the acid distilled. The distillate is tested from time to time with silver nitrate solution, and when a few drops diluted with water fail to become opalescent on the addition of the silver solution, the receiver is changed and the distillation continued until about four-fifths of the acid have passed over. To detect sulphuric acid a small quantity of the acid is treated with a drop of barium chloride and evaporated to dryness in a porcelain basin, water is added and the liquid is poured into a test tube ; the solution should not be in the least degree turbid.

The student will occasionally need to use the red fuming acid obtained by saturating the strongest nitric acid with nitrogen tetroxide. This acid may be prepared by mixing about three parts of starch in small pieces with 200 parts of pure nitre, and pouring over the mixture contained in a large tubulated retort, 100 parts of Nordhausen sulphuric acid previously mixed with 100 parts of the strongest oil of vitriol. The retort should be placed on a sandbath ; the distillation commences without extraneous heat ; if the nitre is not quite free from chlorides the first portions of the distillate must be collected apart. A gentle heat is applied to the retort so soon as the distillation slackens. The operation must be carried on in a draught-chamber or in the open air, and due care must be taken to condense the distillate. The entire process must be conducted at the lowest possible temperature, and not too quickly, or the distillate will be contaminated with sulphuric acid.

3. **Sulphuric acid** : ordinary oil of vitriol almost invariably contains lead, arsenic, iron, and oxides of nitrogen. The readiest method of obtaining it free from these impurities is to heat it with a small quantity of ammonium sulphate and manganese dioxide and distil it. About 700 c.c. of the strong acid are poured into a porcelain dish, from two to three grams of ammonium sulphate are added, and the acid is heated until it fumes strongly ; the oxides of nitrogen are in this manner destroyed. The liquid is allowed to cool, the manganese dioxide added, and the dish again heated nearly to boiling with constant stirring ; the arsenious oxide is thus converted into arsenic pentoxide. The acid is again allowed to cool and transferred to a retort of 1000 c.c. capacity and distilled, a few platinum-clippings being added to moderate the violent *succussive* character of the ebullition. The retort must be supported over the naked flame ; this should be so arranged as to heat the sides of the retort more than the bottom, whereby the tendency to 'bump' is diminished. The neck of the retort may dip directly into the condensing flask ; the latent heat of sulphuric acid vapour is so small that no water is needed to assist the condensation.

4. **Potassium hydrate.** Caustic potash is not readily obtained

Q

pure. The commercial substance may be partially purified by dissolving it in strong alcohol in a stoppered bottle, allowing the insoluble matters to subside, drawing off the clear supernatant dark-brown liquid by means of a syphon, and evaporating it in a silver basin until dense white vapours make their appearance. The resinous matter formed by the action of the alcohol should be skimmed off the surface by a silver spatula or spoon, after which the oily liquid is poured on to plates, where it solidifies; it is quickly broken up and preserved in well-stoppered bottles. Its solution should be kept in bottles free from lead, since the alkali rapidly attacks flint-glass, dissolving out the lead oxide.

The purest form of the hydrate is obtained by igniting a mixture of one part of pure nitre with two parts of thin copper foil cut in small pieces; the mixture is arranged in alternate layers in a covered copper crucible and heated for half an hour to a moderate red heat. When cold the mass is dissolved out by boiling water, and the turbid solution poured into a long stoppered cylinder, in order that the copper oxide may subside. The clear solution is removed by a syphon and rapidly boiled down in a silver basin to the state of the oily hydrate.

Instead of metallic copper ferric oxide may be employed. This is prepared by igniting ferrous oxalate; it is mixed with the nitre in the same proportion as the copper. Through the lid of the crucible runs a tube passing nearly to the bottom, through which a stream of hydrogen is delivered. The crucible is exposed to a low red heat for about an hour; when cold the mass is boiled out with water, and the iron oxide allowed to settle. The clear solution of the hydrate is then evaporated in a silver basin as above described. The ferric oxide may be used repeatedly for the same operation.

5. Sodium hydrate. Very pure caustic soda may be obtained by the action of the metal on water. The solution is prepared by floating a silver dish containing a small quantity of water in a basin of cold water, and adding the metal, cut in very small pieces, to the water in the metallic dish. Care must be taken to allow the liquid to cool after each addition of the metal, or violent explosions may result, and the solution may be projected from the dish. The solution should be prepared when required, if needed in small quantity only; it is not advisable to attempt to manufacture more than a few grams at a time.

Sodium hydrate free from silica and alumina may be obtained by boiling purified sodium carbonate with a little silver carbonate in a silver basin, decanting the clear solution, sufficiently diluting it, and again boiling it, with calcined marble, in the silver dish; the solution is filtered by pouring into a funnel, the stem of which is stopped with pieces of marble, and over which is placed a quantity of pulverised marble. It is advisable before pouring on the alkaline liquid to wash the marble once or twice with distilled water to remove the fine dust.

Impure sodium hydrate may be freed from sulphates and chlorides by exposing a concentrated solution to a low temperature when the hydroxide crystallises out with seven mols. of water. This hydrate melts at six degrees ; by repeated fusions and solidifications it may be obtained perfectly pure.

6. **Sodium carbonate.** The soda of commerce consists principally of the salt $Na_2 CO_3 + 10H_2O$, which requires 62·7 per cent. of water ; in addition it generally contains small quantities of sodium sulphate, chloride sulphide, and thiosulphate, together with traces of calcium and magnesium salts. In order to purify it the salt is recrystallised once or twice from warm solutions, the solution being agitated to prevent the formation of large crystals, which are apt to enclose the impure mother liquors. The crystalline powder is thrown on to a funnel, drained by the action of the pump, and washed with small quantities of ice-cold water. Its solution should give no precipitate with barium chloride after acidification with pure hydrochloric acid, and it should afford only the faintest indication of chlorine when neutralised with pure nitric acid and tested with silver nitrate solution.

7. **Calcium chloride.** The residual solution obtained in the manufacture of carbon dioxide from marble and hydrochloric acid is neutralised, if acid, by the addition of fresh marble in small pieces, and a small quantity of milk of lime is then poured into the solution so as to render it distinctly alkaline. It is heated gently and set aside, in order that the precipitated oxides of iron, magnesia, &c., may subside. The clear liquid is drawn off by means of a syphon, acidified with hydrochloric acid, and evaporated to dryness in a porcelain basin, the residue broken up and heated to about 200° on a sandbath until hydrochloric acid fumes no longer make their appearance. The mass should be broken up into small pieces (about the size of pearl-barley grains) and sifted from the fine powder, which may be worked up on a subsequent occasion. The pieces should be light, porous, and perfectly white. On solution they must not show the least trace of alkalinity.

8. **Soda-lime.** The specific gravity of a strong soda-ley is ascertained by means of the hydrometer, and its strength is calculated from the table given in the 'Manual of Quantitative Analysis.' A definite quantity is then measured off, poured into a clean iron pot, and mixed with small pieces of freshly burnt lime. The weight of lime added should be double that of the soda in the solution. The turbid liquid is evaporated to complete dryness with constant stirring, and the dried mass is

heated to low redness in an iron or clay crucible, and then broken up into small pieces. The soda-lime should be sifted, and the fine and coarse powder preserved separately in well-stoppered bottles. For many purposes a mixture of pure sodium carbonate and calcium hydrate may be used instead of soda-lime.

D.—*Tables.*

I. TABLE OF SOLUBILITIES.—(See Table L. p. 163.)

This table is so arranged that the *bases* are placed at the heads of the columns, the *acids* with which they may be combined, at the side.

The figures refer to the various *menstrua* in which the salts are soluble. Thus :

I.   means soluble in water.
II.   means soluble in acids, but insoluble in water.
III. means insoluble in both water and acids.

The solubilities of the more commonly occurring salts are indicated by large letters 'I. II. III., while those which are of less frequent occurrence have small letters placed after them—1, 2, or 3.

Some substances belong to more than one class—this is indicated thus : 1-2 means a substance difficultly soluble in water, but soluble in acids. 1-3 means a substance soluble with difficulty in water, the solubility of which is not increased by the addition of acids.

2-3 means a substance insoluble in water, and but slightly soluble in acids. The solubilities of the more commonly occurring double salts are given in a separate table. In this table a few of the simple and compound Cyanides are also enumerated.

A small number attached to the figure indicating the solubility of a substance, thus II_6 means that further information is given in the notes to the table.

BASES.

Acids and Halogens, &c.	Group I.			Group II. Div. 1.				Group II. Div. 2. (See Acids)				
	PbO II.	Ag_2O 2	Hg_2O II.	HgO II.	Bi_2O_3 2	CuO II$_2$	CdO 2	As_2O_3 I.-II.	As_2O_5 II.	SnO 2	SnO_2 2-3	Sb_2O_3 2$_3$
S	2	2$_3$	II.$_4$	II.$_4$	2	2	2	II.$_3$	II.$_3$	2$_6$	2$_6$	2$_7$
Cl	I.-III.	III.	II.-III.	I.	1	I.	1			I.	1	I.
I	II.	3	II.				1			2	1	
SO$_4$	II.-III.	I.-III.	1-2	I.	1	I.	I.			1		
NO$_3$	1	1	I.	1	L.$_9$	1	I.					
PO$_4$	2	2	2	2		2	2					
CO$_2$	II.	2	2	2	2	II.	2				2	
BO$_3$	2	1		2	2	2	1-2				2	
C$_2$O$_4$	II.	2	2	2	2	2	2					1-2
ClO$_3$	1	2	1-2	1		1	1					1
As$_2$O$_3$	2	2	2	2	2	2						2
As$_2$O$_5$	2	1	2	2		II.						2
C$_2$H$_3$O$_2$	1	2	1-2	1	1	1^{11}	1			1	1	1
C$_4$H$_4$O$_6$	2	1	1-2	2	2	1	1-2			1-2		I.
CHO$_2$	1	2	1		1	1	1			2		
C$_6$H$_5$O$_7$	2		2			1	1					

BASES. — Continued.

Acids and Halogens, &c.	Group VI.			Group V.				Group IV.				Group III.			
	K₂O I.	Na₂O I.	(NH₄)₂O I.	MgO II.	CaO I.-II.	SrO I.	BaO I.	MnO II.	ZnO II.	NiO II.	CoO II.	Al₂O₃ II.	Cr₂O₃ II.-III.	Fe₂O₃ II.	FeO II.
S	I.	I.	I.	2	I.-II.	I.	I.	II.	II.	II.₈	II.₈	—	—	II.	II.
Cl	I.	I.	I.	1	I.	I.	I.	I.	1	I.	I.	1	I.	I.	I.
I	I.	1	1	1	1	1	1	1	1	—	—	—	—	1	1
SO₄	I.	I.	I.	I.	I.-III.	III.	III.	I.	I.	I.	1	I.	I.	I.	I.
NO₃	I.	I.	I.	1	1	1	1	1	1	1	1	1	1	1	1
PO₄	1	I.	I.	2	II.	2	2	2	2	2	2	2	2	II.	2
CO₂	I.	I.	I.	II.	II.	II.	II.	II.	II.	—	2	—	—	—	2
BO₃	I.	I.	1	2	2	2	2	2	2	2	2	2	2	2	2
C₂O₄	I.	1	I.	2	II.	2	2	2	2	2	2	1	1	2	1-2
ClO₃	I.	1	1	1	1	1	1	1	1	1	1	2	1	1	1
As₂O₃	1	1	I.	2	2	2	2	2	3	2	2	—	—	2	2
As₂O₅	1	1	I.	—	2	1	I.	—	—	2	2	1	1	I.	2
C₂H₃O₂	I.	I.	I.	1	1	1	1	1	1	1	1	1	1	I.	1
C₄H₄O₆	I.	I.	I.	1-2	II.	2	2	1-2	2	—	1	1-2	—	I.	1-2
CHO₂	1	1	1	1	1	1	1	1	1	1	1-2	1-2	—	1	1
C₆H₅O₇	1	1	1	1	2	1	2	1	2	2	—	1-2	1	1	1

TABLE OF SOLUBILITIES.—No. 2.

Cyanide of Potassium	I.
Ferrocyanide ,,	I.
Ferricyanide ,,	I.
Ferrocyanide of Zinc and Potassium	II.
Prussian blue	III.

DOUBLE SALTS.

Potassium-aluminium sulphate	I.
Aluminium-ammonium sulphate	I.
Potassium-ammonium tartar	I.
,, *Sodium* ,,	I.
,, *Ferric* ,,	I.
,, *Antimony* ,,	I.
Sodium-ammonium phosphate	I.
Ammonium-copper chloride	I.
,, *Mercury* ,,	II.
Platinum-ammonium ,,	1–3.
,, *Potassium* ,,	1–3.

Notes to Table of Solubilities.

1. *Hydrochloric acid* converts *minium* into chloride insoluble in excess of the reagent. Nitric acid dissolves it partially, but converts some of it into insoluble brown lead peroxide.

2. *Antimonious oxide*, soluble in hydrochloric, but insoluble in nitric acid.

3. *Silver sulphide*, soluble only in nitric acid.

4. *Mercurous and mercuric sulphides*, soluble only in aqua regia.

5. *Arsenic sulphides*, decomposed slightly by boiling strong hydrochloric acid, but decomposed and dissolved by nitric acid.

6. *Tin sulphides* are decomposed and dissolved by hydrochloric acid, nitric acid converts them into insoluble hydrates.

7. *Antimonious sulphide*, dissolves in strong hydrochloric acid.

8. *Nickel and cobalt sulphides*, much more easily decomposed and dissolved by nitric than by hydrochloric acid.

9. *Basic bismuth nitrate*, soluble in acids.

10. *Basic lead acetate*, partially soluble in water, completely so in acids.

II. Table Showing the Relation between English Weights and Measures and those of the Metric System.

Measures of Weight.

Milligram	=	0·01543235 troy grains.
Centigram	=	0·1543235 ,, ,,
Decigram	=	1·543235 ,, ,,
Gram	=	15·43235 ,, ,,
,,	=	·643 pennyweight.
,,	=	·03216 oz. troy.
,,	=	·03527 oz. avoirdupois.
Kilogram	=	2·6803 lbs. troy.
,,	=	2·20462 lbs. avoirdupois.
Metric Ton (1000 kilos.)	=	2204·62 ,, ,,

Measures of Length.

Millimetre	=	·03937 inches.
Centimetre	=	·3937 ,,
Decimetre	=	3·937 ,,
Metre	=	39·37 ,,
,,	=	3·2809 feet.
,,	=	1·0937 yards.

Inch = 2·53995 centimetres. Foot = 3·04794 decimetres. Yard = ·91438 metres. Mile = 1609·32 metres.

The accompanying scale shows the relation between centimetres and inches.

Inches.

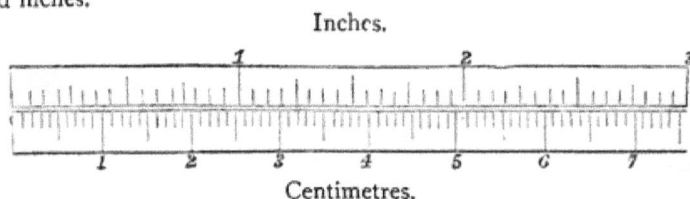

Centimetres.

III. To Convert Degrees Fahrenheit into Degrees Centigrade and vice versa.

$$(C.° \times 1·8) + 32 = F.°$$
$$(F.° - 32) \div 1·8 = C.°$$

IV. To Convert Degrees of Twaddell's Hydrometer into Specific Gravity.

$$(\text{Degrees Twaddell} \times 5) + 1000 = \text{sp. gravity.}$$
$$(\text{Sp. Gravity} - 1000) \div 5 = \text{degrees Twaddell.}$$

V. Names, Symbols, and Atomic Weights of the Elements.

Element	Symbol	Atomic weight	Observer
Aluminium . .	Al	27·26	Berzelius
Antimony .	Sb	122·3	Kessler ; Dexter
Arsenic . .	As	75·15	Kessler
Barium . .	Ba	137·16	Marignac .
Bismuth . .	Bi	210·0	Dumas
Boron . .	Bo	11·04	Berzelius
Bromine .	Br	79·95	Stas
Cadmium . .	Cd	112·04	Lenssen
Cæsium . .	Cs	133·00	Johnson and Allen ; Bunsen
Calcium . .	Ca	40·00	Erdmann and Marchand
Carbon . .	C	12·00	Dumas and Stas ; Liebig
Cerium . .	Ce	92·16	Rammelsberg ; 91·34, Wolf
Chlorine . .	Cl	35·46	Stas
Chromium .	Cr	52·08	Siewert
Cobalt . .	Co	58·74	Russell
Copper . .	Cu	63·12	Millon and Commaille
Didymium .	D	94·96	Hermann .
Erbium . .	E	112·6	Bahr and Bunsen
Fluorine . .	F	18·96	Luca ; Louyet
Glucinum . .	Gl	9·30	Awdejew ; Klatzo
Gold . .	Au	196·71	Berzelius
Hydrogen .	H	1	Dulong and Berzelius
Indium . .	In	113·4	Winkler ; Bunsen
Iodine . .	I	126·85	Stas
Iridium . .	Ir	196·87	Berzelius
Iron . .	Fe	56·00	Dumas
Lanthanum .	La	92·88	Hermann ; 90·18 Zschiesche
Lead . .	Pb	206·92	Stas
Lithium . .	Li	7·02	Stas
Magnesium .	Mg	24·00	Dumas
Manganese .	Mn	54·04	Schneider
Mercury . .	Hg	200·00	Erdmann and Marchand
Molybdenum .	Mo	96·00	Dumas : Debray
Nickel . .	Ni	58·74	Russell
Niobium . .	Nb	94·0	Marignac
Nitrogen . .	N	14·04	Stas
Osmium . .	Os	199·03	Berzelius
Oxygen . .	O	16·00	
Palladium . .	Pd	106·57	Berzelius
Phosphorus .	P	31·00	Schrötter
Platinum . .	Pt	197·18	Andrews
Potassium .	K	39·13	Stas
Rhodium . .	Rh	104·21	Berzelius

V. *Names, Symbols, and Atomic Weights of the Elements.*—Cont.

Element	Symbol	Atomic weight	Observer
Rubidium . .	Rb	85·40	Bunsen ; Piccard
Ruthenium .	Ru	104·40	Berzelius
Selenium . .	Se	79·46	Dumas
Silver . .	Ag	107·93	Stas
Silicon . .	Si	28·10	Dumas
Sodium . .	Na	23·04	Stas
Strontium .	Sr	87·54	Marignac
Sulphur . .	S	32·07	Stas
Tantalum . .	Ta	182·30	Marignac
Tellurium .	Te	128·06	v. Hauer
Thallium . .	Tl	203·64	Crookes
Thorium . .	Th	115·72	Delafontaine
Tin . . .	Sn	118·10	Dumas
Titanium . .	Ti	50·00	Pierre
Tungsten . .	W	184·00	Schneider; Dumas; Roscoe
Uranium . .	U	118·80	Ebelmen
Vanadium .	V	51·35	Roscoe
Yttrium . .	Y	61·70	Bahr and Bunsen
Zinc . .	Zn	65·16	Axel Erdmann
Zirconium .	Zr	89·60	Marignac

INDEX.

ACETIC acid, tests for, 152
Air, analysis of, 29
— combustion of, in coal gas, 16
Alkaloids, detection of, by process of Otto and Stas, 200
— — — — — Uslar and Erdmann, 202
Allotropic, meaning of term, 78
Alum in bread, detection of, 120
Alumina, traces of, detection of, 120
Aluminium, tests for, 120
Ammonia, as group reagent, 119
— combination of, with hydrochloric acid, 34
— — — — — with oxygen, 34
— combustion of, 33
— preparation of, 31
— solubility of, in water, 35
— tests for, 35
Ammonium carbonate, as group reagent, 123
— sulphide, as group reagent, 126
— tests for, 130
Analysis, by dry reactions, 90
— by wet reactions, 90
Analysis, meaning of term, 42
Antimony, in organic mixtures, detection of, 195
— tests for, 115
Apparatus list, 219
Arsenic, in organic mixtures, detection of, 187
— in presence of copper, detection of, 114
— tests for, 113

BARIUM, strontium, and calcium, detection of by flame reactions, 127
Barium, tests for, 126
Bases, reactions of the, 107
Beads, coloured, in Bunsen flame, 100
Benzoic acid, tests for, 153
Bismuth, tests for, 111
Bleaching powder, preparation of, 56
Bloodstains, identification of, 209
Borates, tests for, 137

Border colours, 101
Boric acid, tests for, 137
Boron, in minerals, detection of, 137
Bread, alum in, detection of, 120
Bromides, in presence of iodides and chlorides, detection of, 145
— tests for, 142
Bromine, action of, on phosphorus, 54
— bleaching action of, 54
— detection of, 63
— in organic compounds, detection of, 144
— preparation of, 52
Brucine, in presence of strychnine, detection of, 208
— tests for, 205
Bunsen lamp, structure and flame of, 91
— — use of, 93

CADMIUM, tests for, 113
Cæsium, tests for, 177
Calcium, tests for, 127
— chloride, tests for, 39
— — preparation of pure, 227
Capillary tubes, 96
Calculi, urinary, examination of, 214
Carbon, combination of, with oxygen, 7
— monoxide, preparation of, 40
— — — — from formic acid, 41
— dioxide, decomposition of, by sodium, 38
— — preparation of, 36
— — properties of, 32
— — tests for, 141
Carbonates, tests for, 141
Cerium, tests for, 174
Chemical action, meaning of term, 8
— test, meaning of term, 8
Chlorates, tests for, 151
— in presence of nitrates, detection of, 150
Chloric acid, tests for, 151
Chlorides, in presence of bromides and iodides, detection of, 145
— tests for, 142

Chlorine, action of, on methane, 67
— affinity of, for hydrogen, 45
— as a bleaching agent, 51
— in organic bodies, detection of, 144
— oxidising action of, 46
— oxygen, derivatives of, 49
— preparation of, 43
— tetroxide, — — 52
Citric acid, tests for, 152
— — — in presence of tartaric acid, detection of, 158
Cobalt, detection of, 123
Chromic acid, detection of, 121
Chromium, detection of, 121
Colchicine, detection of, 205
Combustion, meaning of term, 15
Compounds, meaning of term, 89
Conine, tests for, 203
Copper nitrate, preparation of, 30
Copper, in organic mixtures, detection of, 112
— tests for, 112
— — — in presence of arsenic, 114
— traces of, detection of, 112
Corks, boring, 9
Cyanides, analysis of, 146
— in presence of hydrocyanic acid, detection of, 198
— tests for, 145

DIALYSIS, use of, 184
Didymium, test for, 174
Displacement, downward, meaning of term, 37
— upward, meaning of term, 83
Distillation, process of, 20
Dry reactions, analysis by, 90
— — preliminary use of, 106

ELEMENTS, 89
— detection of rare, 170
— number, 90
— rare, systematic search for, 178
Emission of light, 94
Ethene, combination of, with bromine and chlorine, 69
— forms explosive mixture with oxygen, 69
— preparation of, 68
Ethine, detection of, 72
— in coal gas, 72
— preparation of, 71

FAHL ore, analysis of, 130
Ferrous sulphate, as test for nitric acid, 22
Films on porcelain, 98
Filters, how to make, 62
Filtrate, meaning of term, 62
Filtration, 62
Flame, colours, 101
— luminous and non-luminous, 73
— reactions, 90

Flame, structure of, 73
— zones in, 75
Flasks, making three-necked, 16
Fluorides, test for, 138
Formic acid, tests for, 154
Fresenius and von Babo's test for arsenic, 192

GAS bottles, 5
Glass tube bending, 3
— — sealing of, 80
Gold, tests for, 171
Group reagent, meaning of term, 107
Groups, number, and what they include, 108
— Group I., separation of, 110
— — II., separation of, 117
— — III., separation of, 122
— — IV., separation of, 125
— — V., separation of, 117
— — VI., separation of, 130
Guaiacum, as test for hydrocyanic acid, 195

HYDRIODIC acid, preparation of, 57
— — properties of, 58
— — tests for, 143
Hydrobromic acid, 58
— tests for, 142
Hydrochloric acid, as group reagent, 108
— — preparation of, 47
— — pure, 224
— — solubility of, in water, 47
— — synthesis of, 48
— — tests for, 142
Hydrocyanic acid, in organic mixtures, detection of, 197
— — tests for, 145
Hydrofluoric acid, action of on glass, 60
— — preparation of, 59
— — tests for, 138
Hydrofluosilicic acid, preparation of, 63
— — tests for, 135
Hydrogen, a combustible body, 11
— burning of oxygen in, 15
— explosion of with oxygen, 14
— lightness of, 11
— preparation of, 9
Hydrogen, test for purity of, 10
Hydrosulphuric acid, tests for, 148
Hypochlorites, 147
Hypochlorous acid, 147

INDIUM, tests for, 176
Inorganic poisons, in organic mixtures, detection of, 183
Iodates, tests for, 141
Iodic acid, 141
Iodides, in presence of chlorides and bromides, detection of, 145
— tests for, 143
Iodide films, 99

Iodine, action of on phosphorus, 56
Iodine, action of on sodium, 56
— — — — on starch, 56
— in organic compounds, detection of, 144
— preparation of, 55
Iron, tests for, 119

LANTHANUM, tests for, 174
Lead, detection of in organic mixtures, 195
— tests for, 110
Light, emission of, 94
Lithium, tests for, 177

MAGNESIUM, tests for, 128
Manganese dioxide, use of in making oxygen, 3
— tests for, 125
Mantle colours, 101
Marsh's process for detection of arsenic, 187
Meconic acid, detection of, 208
Melting points, 94
Mercury (monad), tests for, 109
— (dyad), detection of in organic mixtures, 195
Metaphosphoric acid, tests for, 136
Methane, combination of with chlorine, 67
— — — with oxygen, 66
— forms an explosive mixture with oxygen, 66
— preparation of, 65
Molybdenum, tests for, 173
Morphine, tests for, 206

NARCOTINE, tests for, 205
Neutralisation, meaning of term, 42
Nickel, tests for, 122
Nicotine, tests for, 203
Niobium, tests for, 171
Nitrates, 149
— and nitrites, discrimination between, 147
— in presence of chlorates, detection of, 150
Nitric acid, preparation of, 19
— — pure, preparation of, 224
— tests for, 149
Nitrites, 147
Nitrogen, in organic bodies, detection of, 151
— preparation of, 18
— properties of, 19
— dioxide, combination of with oxygen, 28
— — combustion in, 28
— — preparation of, 27
— monoxide, preparation of, 23
— tetroxide, production of in testing for nitric acid, 21
Nitrous acid, tests for, 147

OPIUM, detection of, 208
Organic poisons, detection of, 196
Orthophosphoric acid, tests for, 135
Osmium, tests for, 172
Oxalates, tests for, 137
Oxalic acid, in presence of organic matter, detection of, 199
— — tests for, 137
Oxygen derivatives of chlorine, 49
—, combination of with hydrogen, 12
—, burning of in hydrogen, 15
— supports combustion, 6
— tests for, 5

PALLADIUM, tests for, 171
Perchlorates, tests for, 51
Perchloric acid, 151
Permanent gas, meaning of term, 85
Phosphates, tests for, 135
—, traces of, detection of, 136
Phosphoretted hydrogen, preparation of, 88
Phosphoric acid, tests for, 135
Phosphorus, action of on bromine, 54
— — — on iodine, 56
— — — on nitrogen monoxide, 25
— detection of, 182
— withdrawal of oxygen from air by, 18
Picoline, tests for, 203
Platinum, tests for, 171
Platinum ores, analysis of, 171
Pneumatic trough, use of, 4
Poisons, detection of, 181
Potassium hydrate, preparation of pure, 225
— tests for, 129
— spectrum of, 129
Potassium chlorate, preparation of, 51
— cyanide, purification of from cyanate, 123
Precipitate, meaning of term, 2
Pyrophosphoric acid, tests for, 136

QUALITATIVE analysis, 89
— — subdivision of, 90
Quinine, sulphate, examination of, 207
— tests for, 207

RARE elements, detection of, 170
— — systematic search for, 178
Reagent list, 220
Reagents, pure, preparation of, 224
Reduction of substances in Bunsen flame, 95
Rhodium, tests for, 172
Rubidium, tests for, 177
Ruthenium, tests for, 171

SEDIMENTS, urinary, examination of, 213

Selenium, tests for, 171
Silicates, analysis of, 140
Silicic acid, tests for, 139
Silico-fluorides, tests for, 135
Silicon fluoride, action of on water, 60
— — preparation of, 60
Silver, tests for, 108
Soda lime, preparation of, 227
Sodium carbonate, preparation of pure, 227
— flame, reaction of, 129
— hydrate, preparation of pure, 226
— spectrum of, 129
— tests for, 129
Solubilities, table of, 228
Spectroscope, direct vision, 105
— use of the, 102
Strontium, tests for, 126
Strychnine, in beer, detection of, 207
— tests for, 206
Sulphates, tests for, 134
Sulphides, in presence of sulphites and thiosulphates, detection of, 148
— tests for, 148
Sulphide films, 100
Sulphites, in presence of sulphides and thiosulphates, detection of, 148
— tests for, 139
Sulphur, combination of with oxygen, 7
— detection of, 78
— properties of, 77
— traces of, detection of, 134
Sulphur dioxide, bleaching action of, 81
— — detection of, 139
— — liquefaction of, 80
— — preparation of, 79
Sulphur trioxide, preparation of, 84
Sulphuretted hydrogen, action of on metallic solutions, 87
— — as group reagent, 111
— — preparation of, 86
Sulphuric acid, free, detection of, 134
— — properties of, 85
— — pure, preparation of, 225
— — tests for, 134
Sulphurous acid, free, detection of, 139
— — preparation of, 84
— — tests for, 139
Synthesis, meaning of term, 42

TABLE of solubilities, 228
— for examination by Bunsen flame, 156

Table for examination by heating in glass tube, 155
— — — of insoluble substances, 166
— for preliminary examination for acids, 165
— for separation of inorganic acids, 167
— — — organic acids, 169
— — solution of substances, 157
— — treatment of Group I., 159
— — — — — II., 160
— — — — — III., 161
— — — — — IV., 162
— — — — — V., 163
— — — — — VI., 164
— — — of solutions, 158
Tables for weights and measures, 232
Tartaric acid, in presence of citr acid, detection of, 153
— — tests for, 152
Tellurium, tests for, 173
Thallium, tests for, 170
Thiosulphates, in presence of sulphides and sulphites, detection of, 148
— tests for, 141
Tin, tests for, 116
Titanium, tests for, 175
Tungsten, tests for, 170

URANIUM, tests for, 176
Urea, tests for, 212
Uric acid, tests for, 212
Urine, examination of, 210
Urinary calculi, examination of, 214
Urinary sediments, examination of, 213

VANADIUM, tests for, 177
Veratrine, tests for, 204

WASH-BOTTLE, making a, 64
Water, composed of, 14
Wet reaction, analysis by, 90

ZINC, detection of, in organic mixtures, 183
— action of on sulphuric acid, 10
— tests for, 124
— sulphate, preparation of, 12
Zirconium, tests for, 174

LONDON: PRINTED BY
SPOTTISWOODE AND CO., NEW-STREET SQUARE
AND PARLIAMENT STREET

MESSRS. LONGMAN & CO.'S TEXT-BOOKS OF SCIENCE, MECHANICAL AND PHYSICAL,
ADAPTED FOR THE USE OF ARTISANS AND OF STUDENTS
IN PUBLIC AND SCIENCE SCHOOLS.

*Now in course of publication, in small 8vo. each volume containing
about Three Hundred pages,*

A SERIES OF
ELEMENTARY WORKS ON MECHANICAL AND PHYSICAL SCIENCE,
FORMING A SERIES OF
TEXT-BOOKS OF SCIENCE
ADAPTED FOR THE USE OF ARTISANS AND OF STUDENTS IN PUBLIC AND
OTHER SCHOOLS.

The first Thirteen of the Series edited by T. M. GOODEVE, M.A. Barrister-at-Law, Lecturer on Applied Mechanics at the Royal School of Mines; and the remainder by C. W. MERRIFIELD, F.R.S. an Examiner in the Department of Public Education, and late Principal of the Royal School of Naval Architecture and Marine Engineering, South Kensington.

THE FOLLOWING, NOW PUBLISHED, ARE EDITED BY T. M. GOODEVE, M.A.

THE ELEMENTS OF MECHANISM.
Designed for Students of Applied Mechanics. By T. M. GOODEVE, M.A. Barrister-at-Law, Lecturer on Mechanics at the Royal School of Mines. New Edition, revised; with 257 Figures on Wood. Price 3s. 6d.

METALS, THEIR PROPERTIES AND TREATMENT.
By CHARLES LOUDON BLOXAM, Professor of Chemistry in King's College, London; Professor of Chemistry in the Department of Artillery Studies, and in the Royal Military Academy, Woolwich. With 105 Figures on Wood. Price 3s. 6d.

INTRODUCTION TO THE STUDY OF INORGANIC CHEMISTRY.
By WILLIAM ALLEN MILLER, M.D. LL.D. F.R.S. late Professor of Chemistry in King's College, London; Author of 'Elements of Chemistry, Theoretical and Practical.' New Edition, revised; with 71 Figures on Wood. Price 3s. 6d.

ALGEBRA AND TRIGONOMETRY.
By the Rev. WILLIAM NATHANIEL GRIFFIN, B.D. sometime Fellow of St. John's College, Cambridge. Price 3s. 6d.

NOTES ON THE ELEMENTS OF ALGEBRA AND TRIGONOMETRY:
With SOLUTIONS of the more difficult QUESTIONS. By the Rev. WILLIAM NATHANIEL GRIFFIN, B.D. sometime Fellow of St. John's College, Cambridge. Price 3s. 6d.

PLANE AND SOLID GEOMETRY.
By the Rev. H. W. WATSON, formerly Fellow of Trinity College, Cambridge, and late Assistant-Master of Harrow School. Price 3s. 6d.

THEORY OF HEAT.
By J. CLERK MAXWELL, M.A. LL.D. Edin. F.R.SS. L. & E. Professor of Experimental Physics in the University of Cambridge. New Edition, revised; with 41 Woodcuts and Diagrams. Price 3s. 6d.

TECHNICAL ARITHMETIC AND MENSURATION.
By CHARLES W. MERRIFIELD, F.R.S. an Examiner in the Department of Public Education, and late Principal of the Royal School of Naval Architecture and Marine Engineering, South Kensington. Price 3s. 6d.

KEY TO MERRIFIELD'S TEXT-BOOK OF TECHNICAL ARITHMETIC AND MENSURATION.
By the Rev. JOHN HUNTER, M.A. one of the National Society's Examiners of Middle-Class Schools; formerly Vice-Principal of the National Society's Training College, Battersea. Price 3s. 6d.

ON THE STRENGTH OF MATERIALS AND STRUCTURES:

The Strength of Materials as depending on their quality and as ascertained by Testing Apparatus ; the Strength of Structures, as depending on their form and arrangement, and on the materials of which they are composed. By JOHN ANDERSON, C.E., LL.D. F.R.S.E. Superintendent of Machinery to the War Department. Price 3s. 6d.

ELECTRICITY AND MAGNETISM.

By FLEEMING JENKIN, F.R.SS. L. & E. Professor of Engineering in the University of Edinburgh. New Edition, revised. Price 3s. 6d.

WORKSHOP APPLIANCES.

Including Descriptions of the Gauging and Measuring Instruments, the Hand Cutting-Tools, Lathes, Drilling, Planing, and other Machine Tools used by Engineers. By C. P. B. SHELLEY, Civil Engineer, Hon. Fellow and Professor of Manufacturing Art and Machinery at King's College, London. With 209 Figures on Wood. Price 3s. 6d.

PRINCIPLES OF MECHANICS.

By T. M. GOODEVE, M.A. Barrister-at-Law, Lecturer on Applied Mechanics at the Royal School of Mines. Price 3s. 6d.

THE FOLLOWING, NOW PUBLISHED, ARE EDITED BY C. W. MERRIFIELD, F.R.S.

QUANTITATIVE CHEMICAL ANALYSIS.

By T. E. THORPE, F.R.S.E. Ph.D. Professor of Chemistry in the Andersonian University, Glasgow. With 88 Figures on Wood. Price 4s. 6d.

QUALITATIVE ANALYSIS AND LABORATORY PRACTICE.

By T. E. THORPE, Ph.D. F.R.S.E. Professor of Chemistry in the Andersonian University, Glasgow ; and M. M. PATTISON MUIR. Price 3s. 6d.

INTRODUCTION TO THE STUDY OF ORGANIC CHEMISTRY;

The CHEMISTRY of CARBON and its COMPOUNDS. By HENRY E. ARMSTRONG, Ph.D. F.C.S. Professor of Chemistry in the London Institution. With 8 Figures on Wood. Price 3s. 6d.

Text-Books *preparing for publication:—*

THE FOLLOWING TO BE EDITED BY T. M. GOODEVE, M.A.

ECONOMICAL APPLICATIONS OF HEAT.

Including Combustion, Evaporation, Furnaces, Flues, and Boilers. By C. P. B. SHELLEY, Civil Engineer, and Professor of Manufacturing Art and Machinery at King's College, London. With a Chapter on the Probable Future Development of the Science of Heat, by C. WILLIAM SIEMENS, F.R.S.

THE STEAM ENGINE.

By T. M. GOODEVE, M.A. Barrister-at-Law, Lecturer on Mechanics at the Royal School of Mines.

SOUND AND LIGHT.

By G. G. STOKES, M.A. D.C.L. Fellow of Pembroke College, Cambridge ; Lucasian Professor of Mathematics in the University of Cambridge ; and Secretary to the Royal Society.

THE FOLLOWING TO BE EDITED BY C. W. MERRIFIELD, F.R.S.

TELEGRAPHY.

By W. H. PREECE, C.E. Divisional Engineer, Post-Office Telegraphs ; and J. SIVEWRIGHT, M.A. Superintendent (Engineering Department) Post-Office Telegraphs.

PRACTICAL AND DESCRIPTIVE GEOMETRY, AND PRINCIPLES OF MECHANICAL DRAWING.

By C. W. MERRIFIELD, F.R.S. an Examiner in the Department of Public Education, and late Principal of the Royal School of Naval Architecture and Marine Engineering, South Kensington.

ELEMENTS OF MACHINE DESIGN.

With Rules and Tables for Designing and Drawing the Details of Machinery. Adapted to the use of Mechanical Draughtsmen and Teachers of Machine Drawing. By W. CAWTHORNE UNWIN, B.Sc. Assoc. Inst. C.E. Professor of Hydraulic and Mechanical Engineering at Cooper's Hill College.

PHYSICAL GEOGRAPHY.

By the Rev. GEORGE BUTLER, M.A. Principal of Liverpool College ; Editor of 'The Public Schools Atlas of Modern Geography.'

London : LONGMANS & CO.